Plant Biotechnology

Poonam Pandey

AMIGA PRESS. INC
Delhi -110051

AMIGA PRESS. INC
H.O :- 28-A, Indra Park, Chander Nagar
Delhi :- 110051 (India)
Tel :- 011-22504028, +91-98736-57000
Email :- amigapress@gmail.com

This edition is published by Amiga Press.Inc

ISBN : 978-93-84533-15-1

First Published 2018

Printed and bound in India by Thomson Press Ltd., New Delhi

Preface

Plant biotechnology is a discipline that should be well known and accepted as a sub field under the umbrella of pharmaceutical biotechnology. Today, biotechnology is being used as a tool to give plants new traits that benefit agricultural production, the environment, and human nutrition and health. The purpose of this publication is to provide basic information about plant biotechnology and to give examples of its uses. Modern biotechnology is purported to have a number of products for addressing certain food-security problems of developing countries. It offers the possibility of an agricultural system that is more reliant on biological processes rather than chemical applications.

The constant aim of plant breeding is to improve the quality, diversity and performance of agricultural and horticultural crops. The overriding objective is to develop plants better adapted to human needs. The evolution of plant breeding is a classic example of how improved biological understanding has been adapted to provide more effective methods of meeting the demands of a changing world. Plant breeding has existed in its most primitive form since the first farmers saved the seeds of their best plants from one season to the next more than 10,000 years ago. Over the centuries, this selection process has gradually become more scientific, bringing major improvements in the yield, quality and diversity of crops grown worldwide.

Modern biotechnology is purported to have a number of products for addressing certain food-security problems of developing countries. It offers the possibility of an agricultural system that is more reliant on biological processes rather than chemical applications.

The potential uses of modern biotechnology in agriculture include: increasing yields while reducing inputs of fertilizers, herbicides and insecticides; conferring drought or salt tolerance on crop plants; increasing shelf-life; reducing post harvest losses; increasing the nutrient content of produce; and delivering vaccines. The availability of such products could not only have an important role in reducing hunger and increasing food security, but also have the potential to address some of the health problems of the developing world.

Achieving the improvements in crop yields expected in developing countries can help to alleviate poverty: directly by increasing the household incomes of small farmers who adopt these technologies; and indirectly, through their positive impact as evidenced in the price slumps of herbicides and insecticides.

Indeed, some developing countries have identified priority areas such as tolerances to alkaline earth metals, drought and soil salinity, disease resistance, crop yields and nutritionally enhanced crops. The adoption of technologies designed to prolong shelf-life could be valuable in helping to reduce post harvest losses in regionally important crops. Prime candidates in terms of crops of choice for development are the so-called 'orphan crops', such as cassava, sweet potato, millet, sorghum and yam.

Currently, the many promises of modern biotechnology that could have an impact on food security have not been realized in most developing countries. In fact, the uptake of modern biotechnology has been remarkably low owing to the number of factors that underpin food security issues. In part, this could be because the first generation of commercially available crops using modern biotechnology were modified with single genes to impart resistance to pests, weed and insects, and not complex characteristics that would modify the growth of crops in harsh conditions. Secondly, the technologies are developed by companies in industrialized countries with little or no direct investment in, and which derive little economic benefit from, developing countries.

This book provides an authoritative review of many aspects of current interest and progress in the field of plant biotechnology that has been made in the recent past.

—*Editor*

Contents

Preface (*v*)

1. **Biotechnology and Plants** 1
The Science of Modern Plant Biotechnology; Biotech Basics; National Biotechnology Policy-Salient Features; Rising Food Prices; Science and Technology; Resource Alienation and Commodification of Common Resources; Globalisation and the Marginalisation of the Farmer; Our Demands; Biotechnology: Indian Scenario; Vaccines; Medicinal Discovery; Agriculture Sector; Medicinal Research; Clinical Research and Trials; Bioinformatics; Modern biotechnology-Sustainable Growth

2. **Plant Genetic Engineering** 18
Plant Genetics; Understanding Gene Testing; RNA and DNA Revealed: New Roles, New Rules; RNA Interference (RNAi); Biological Cell; Biological Energy-ADP and ATP; Cell Respiration; Biological Viruses; The Tools of Genetics: Recombinant DNA and Cloning; Discovery of the Hybrid Helix and the First DNA-RNA Hybridization ; A New Type of Polymer Chemistry

3. **Gene Mapping of Plant** 50
Transposition; Amplification And Rearrangement; Genome Contraction; Controlling Transcription, Recombination, And Transposition; Homology-dependent Gene Silencing; The Origin Of Transposons And Methylation; Transposons And Genome Evolution In Plants; The Discovery Of Transposition; Plant Transposons In The Age Of Genomics; The Paradox; Synteny And Divergence

4. **Plant Cell Structure** 77
History of Plant Cell; Pioneering Pathfinders; A. C. Seward and the Cryptozoon Controversy; Emergence Of A New Field Of Science; Lessons From The Hunt; Crystalline Inclusions within Plant Cells; Eukaryotic Cells and Plant Cell; Anatomy of Plant Cell; Early Evolution and the Origin of Cells; An Appreciation; Solution to Darwin's Dilemma

5. **Plant Reproduction** **105**
Reproduction in Plants; Reproductive Systems and Evolution in Vascular Plants; Modes Of Reproduction; Consequences Of Reproductive Systems; Evolution Of Reproductive Systems; The Cost Of Sex; Plant Reproduction System; Genetic Variations in Plants ; Describing and Observing Phenotypic Variation ; Genetic Diversity; Diagram of a Dicot Plant; Importance of Mode of Reproduction; Self Fertilizing Crops (Autogamous Crops); Selection Methods for the Development of Pure Breeding Cultivars from Crosses; Factors affecting plant reproduction; Conclusions

6. **Transgenic Plants for Recombinant Proteins** **149**
Legitimate Concerns about Transgenic Ariculture; Misinformation about Food Safety; Some Early Fruits of Transgenic Agriculture; Transgenic Plants for Antibodies ; Transgenic Plants with Beneficial Traits; Production of Polyhydroxybutyrate or PHB

7. **Plant Breeding** **192**
Introduction; Plant Breeding For A World Community; Speeding-up Plant-breeding Work; Breeding Methods in Crop Plants; Cross Pollinated Crops; Mutation Breeding; Chemical Mutagens; Participatory Plant Breeding; Conventional Breeding; Plant Breeding in the Future

Bibliography **221**

Index **223**

1

Biotechnology and Plants

Today, biotechnology is being used as a tool to give plants new traits that benefit agricultural production, the environment, and human nutrition and health. The purpose of this publication is to provide basic information about plant biotechnology and to give examples of its uses.

The goal of plant breeding is to combine desirable traits from different varieties of plants to produce plants of superior quality.

This approach to improving crop production has been very successful over the years. For example, it would be beneficial to cross a tomato plant that bears sweeter fruit with one that exhibits increased disease resistance.

To do this, it takes many years of crossing and backcrossing generations of plants to obtain the desired trait. Along the way, undesirable traits may be manifested in the plants because there is no way to select for one trait without affecting others.

Another limitation of traditional plant selection is that breeding is restricted to plants that can sexually mate.

Advances in scientific discovery and laboratory techniques during the last half of the twentieth century led to the ability to manipulate the deoxyribonucleic acid (DNA) of organisms, which accelerated the process of plant improvement through the use of biotechnology.

THE SCIENCE OF MODERN PLANT BIOTECHNOLOGY

Genes and the Genome

Plants are made of millions of cells all working together. Every cell of a plant has a complete "instruction manual" or genome (pronounced "JEE-nom") that is inherited from the parents of the plant as a combination of their genomes.

Genes are found within the genome and serve as the "words" of the instruction manual.

When a cell reads a word, or in scientific terms "expresses a gene," a specific protein is produced. Proteins give an individual cell, and therefore the plant, its form and function. Genes (words) are written using the four-letter alphabet A, C, G, T. The letters are abbreviations for four chemicals called bases, which together make up DNA. DNA is universal in nature, meaning that the four chemical bases of DNA are the same in all living organisms. Consequently, a gene from one organism can function in any other organism.

The ability to move genes into plants from other organisms, thereby producing new proteins in the plant, has resulted in significant achievements in plant biotechnology that were not possible using traditional breeding practices.

Methods of Introducing Genes into Plants

To genetically modify a plant, the thousands of bases of DNA comprising an individual gene are transferred into an individual plant cell where the new gene becomes a permanent part of the cell's genome. This process makes the resulting plant "transgenic." Transfer of DNA into plant cells is done using various "transformation" techniques that are the result of discoveries in basic Science.

Nature's way

One method to transfer DNA into plants takes advantage of a system found in nature. The bacterium that causes "crown gall tumors" injects its DNA into a plant genome, forcing the plant to create a suitable environment for the bacterium to live. After discovering this process, scientists were able to "disarm" the bacterium, put new genes into it, and use the bacterium to harmlessly insert the desired genes into the plant genome.

Cellular Target Practice

In the "biolistic" or "gene gun" method, microscopic gold beads are coated with the gene of interest and shot into the plant cell with a burst of helium. Once inside the cell, the gene comes off the bead and integrates into the cell's genome.

That's Shocking!

It was also discovered that plant cells could be "electroporated" or mixed with a gene and "shocked" with a pulse of electricity, causing holes to form in the cell through which the DNA could flow. The cell is subsequently able to repair the holes and the gene becomes a part of the plant genome.

Selecting the Right Cells

When using these methods, new genes are successfully introduced into only a small percentage of the cells, so scientists must be able to "pick out" or "select" the transformed cells before proceeding. This is often done by concurrently introducing an additional gene into the cell that will make it resistant to an antibiotic.

A cell that survives antibiotic treatment will most likely have received the gene of interest as well; that cell is subsequently used to propagate the new plant. There is a concern that the gene giving antibiotic resistance could naturally be transferred to bacteria once the transgenic plant is in the wild, making bacteria resistant to antibiotics that are used to fight human infection. Scientists are currently devising ways to select for transformed cells that will alleviate this issue.

Traits Being Introduced Into Plants

Changes made to plants through the use of biotechnology can be categorized into the three broad areas of input, output, and value-added traits. Examples of each are described below.

Input Traits

An "input" trait helps producers by lowering the cost of production, improving crop yields, and reducing the level of chemicals required for the control of insects, diseases, and weeds.

Input traits that are commercially available or being tested in plants:

- Resistance to destruction by insects
- Tolerance to broad-spectrum herbicides
- Resistance to diseases caused by viruses, bacteria, fungi, and worms
- Protection from environmental stresses such as heat, cold, drought, and high salt concentration

Output Traits

An "output" trait helps consumers by enhancing the quality of the food and fiber products they use.

Output traits that consumers may one day be able to take advantage of:

- Nutritionally enhanced foods that contain more starch or protein, more vitamins, more anti-oxidants (to reduce the risk of certain cancers), and fewer trans-fatty acids (to lower the risk of heart disease)
- Foods with improved taste, increased shelf-life, and better ripening characteristics
- Trees that make it possible to produce paper with less environmental damage

- Nicotine-free tobacco
- Ornamental flowers with new colors, fragrances, and increased longevity

"Value-added" Traits

Genes are being placed into plants that completely change the way they are used.

Plants may be used as "manufacturing facilities" to inexpensively produce large quantities of materials including:

- Therapeutic proteins for disease treatment and vaccination
- Textile fibers
- Biodegradable plastics
- Oils for use in paints, detergents, and lubricants

Plants are being produced with entirely new functions that enable them to do things such as:

- Detect and/or dispose of environmental contaminants like mercury, lead, and petroleum products

Canola Plants made Resistant to High Concentrations of Salt Through Biotechnology

Canola plants grown in the presence of a high concentration of salt. Non-genetically modified canola (non-GM) or canola genetically modified to have high, medium, or low tolerance to salt.

Plants with "Input Traits" that are Commercially available include

- Roundup Ready soybean, canola, and corn: resistant to treatment with Roundup herbicide that may result in more effective weed control with less tillage, and/or decreased use of other, more harmful herbicides
- YieldGard corn and Bollgard cotton: express an insecticidal protein that is not toxic to animals or humans which protects the plant from damage caused by the European corn borer, tobacco budworm, and bollworm
- Destiny III and Liberator III squash: resistant to some viruses that destroy squash

Plants may become available with "Output Traits" including

- High laurate canola and high oleic soybean having altered oil content to be used primarily in industrial oils and fluids rather than food
- High-starch potatoes that take up less oil when frying
- Longer shelf-life bananas, peppers, pineapples, strawberries, and tomatoes

- Soybeans with higher levels of isoflavones; compounds that may be beneficial in reducing some cancers and heart disease
- Plants that produce vaccines and pharmaceuticals for treatment of human diseases
- Corn with improved digestibility and more nutrients providing livestock with better feed

Issues Associated with Genetically Modified Plants

Benefits and Risks

The list of plants and plant-derived products made as a result of modern biotechnology is everincreasing. Many transgenic plants, such as herbicide resistant soybeans, have been widely adopted by producers signifying their satisfaction, while other products, such as the delayed softening "FlavrSavr" tomato, are no longer on the market. Some of the potential benefits from using transgenic plants include:

- Reduced crop production costs and increased yields
- Healthier, more nutritious foods
- Reduced environmental impact from farming and industry
- Increased food availability for underdeveloped countries

Potential risks associated with transgenic plants include:

- Introduction of allergenic or otherwise harmful proteins into foods
- Transfer of transgenic properties to viruses, bacteria, or other plants
- Detrimental effects on non-target species and the environment

Safety, Regulation, and Labeling

In the United States

At the Federal level, the Food and Drug Administration, the Environmental Protection Agency, and the Department of Agriculture extensively review products of biotechnology to ensure that they are safe for public use and the environment.

GM foods require labeling only if they differ significantly in safety, composition, or nutritional content when compared to their non-GM counterparts. Additionally, the FDA requires a GM food to be labeled if it contains a known allergen unless data have shown that there is no allergy risk.

In Organic Products

Organic standards reflect a "zero tolerance" policy concerning transgenic products and organisms. Organic food producers are taking precautions to minimize the risk of unintentional contamination of their products with transgenic ones.

In Canada

The Canadian Food Inspection Agency, Health Canada, and Environment Canada strictly regulate agricultural biotechnology products. They currently require GM foods to be labeled if they differ significantly in composition or nutritional value and support a voluntary labeling policy for others.

In Europe

The acceptance of GM crops by the European Union has been more reserved. However, recent statements made by European Union officials suggest that their position may be changing as they are calling for their policies regarding GM crops to be based on scientific principles rather than on public opinion and misconceptions. Europe currently favors labeling of all GM foods and a system that would allow for "identity preserved" processing in which foods would be guaranteed to contain no genetically modified products.

(credit: USDA National Agricultural Statistics Service, Pew Initiative on Food and Biotechnology)

BIOTECH BASICS

What has come to be called "biotechnology" and the genetic enhancement of agricultural products may be one of the oldest human activities. For thousands of years, from the time human communities began to settle in one place, cultivate crops and farm the land, humans have manipulated the genetic nature of the crops and animals they raise. Crops have been bred to improve yields, enhance taste and extend the growing season.

Each of the 15 major crop plants, which provide 90 percent of the globe's food and energy intake, has been extensively manipulated, hybridized, inter-bred and modified over the millennia by countless generations of farmers intent on producing crops in the most effective and efficient ways possible.

Today, biotechnology holds out promise for consumers seeking quality, safety and taste in their food choices; for farmers seeking new methods to improve their productivity and profitability; and for governments and non-governmental public advocates seeking to stave off global hunger, assure environmental quality, preserve bio-diversity and promote health and food safety.

The Basics of Plant Biotechnology

For centuries, humankind has made improvements to crop plants through selective breeding and hybridization — the controlled pollination of plants.

Plant biotechnology is an extension of this traditional plant breeding with one very important difference – plant biotechnology allows for the transfer of a greater variety of genetic information in a more precise, controlled manner.

Traditional plant breeding involves the crossing of hundreds or thousands of genes, whereas plant biotechnology allows for the transfer of only one or a few desirable genes. This more precise Science allows plant breeders to develop crops with specific beneficial traits and without undesirable traits.

Many of these beneficial traits in new plant varieties fight plant pests – insects, weeds and diseases – that can be devastating to crops. Others provide quality improvements, such as tastier fruits and vegetables; processing advantages, such as tomatoes with higher solids content; and nutrition enhancements, such as oil seeds that produce oils with lower saturated fat content. Crop improvements like these can help provide an abundant, healthful food supply and protect our environment for future generations.

"Modern techniques of genetic engineering are essentially a refinement of the kinds of genetic modification that have long been used to enhance plants, micro-organisms, and animals for food. The products of the newer techniques are even more predictable and safer than the genetically engineered foods that have long enriched our diet."

How Biotechnology Works

The DNA (deoxyribonucleic acid) from different organisms is essentially the same – simply a set of instructions that directs cells to make the proteins that are the basis of life. Whether the DNA is from a microorganism, a plant, an animal or a human, it is made from the same materials.

Throughout the years, researchers have discovered how to transfer a specific piece of DNA from one organism to another.

A researcher's first step in transferring DNA is to "cut" or remove a gene segment from a chain of DNA using enzyme "scissors" to cut at a specific site along the DNA strand.

The researcher then uses these "scissors" to cut an opening into the plasmid – the ring of DNA often found in bacteria outside of a cell. Next, the researcher "pastes" or places the gene segment into the plasmid. Because the cut ends of both the plasmid and the gene segment are chemically "sticky," they attach to each other, forming a plasmid containing the new gene. To complete the process, researchers use another enzyme to paste or secure the new gene in place.

Decades of research have allowed Monsanto specialists to apply their knowledge of genetics to improve various crops, such as corn, soybeans, canola, cotton and potatoes.

Our researchers continue to work carefully to ensure that improved crops are the same as current crops, except for the addition of one beneficial trait which, for example, protects them from a particular insect or virus.

"...to maintain the productivity of agriculture, we must continue to improve the agricultural seeds that are used... We are now blessed through research and technology with new methods of actually speeding up the process of improving the seeds and the products we get from them...

Why Biotechnology Matters

Many people are beginning to appreciate more deeply the bonds between human well-being, social stability and the natural processes of the earth that sustain all life. They are realizing that the earth's capacity to continue providing clean air and water, productive soils and a rich diversity of plant and animal life is central to ensuring quality of life for ourselves and our descendants.

Current population growth is already straining the earth's resources. One of the few certainties of the future is that the world's population will nearly double, reaching almost 10 billion inhabitants by the year 2030. According to Ismail Serageldin of the World Bank, "Biotechnology will be a crucial part of expanding agricultural productivity in the 21st century. If safely deployed, it could be a tremendous help in meeting the challenge of feeding an additional three billion human beings, 95% of them in the poor developing countries, on the same amount of land and water currently available."

Biotechnology, which allows the transfer of a gene for a specific trait from one plant variety or species to another, is one important piece of the puzzle of sustainable development.

Experts assert that biotechnology innovations will triple crop yields without requiring any additional farmland, saving valuable rain forests and animal habitats. Other innovations can reduce or eliminate reliance on pesticides and herbicides that may contribute to environmental degradation. Still others can preserve precious groundsoils and water resources.

Most experts agree that the world doesn't have the luxury of waiting to act. By working now to put in place the technology and the infrastructure required to meet future food needs, we can feed the world for centuries to come and improve the quality of life for people worldwide.

"The possibility that (biotech) crops could make a substantial contribution to providing sufficient food for an expanding world is, on its own, a solid reason for engaging in the research that underlies their development."

Benefits of Biotechnology

Now and in the near future, the products of food biotechnology provide food quality improvements which include better taste and healthier foods.

Agronomic or "input" traits create value by giving plants the ability to do things that increase production or reduce the need for other inputs such as chemical pesticides or fertilizers. Our current products with input traits include potatoes, corn and soybeans that produce better yields with fewer costly inputs through better control of pests and weeds. Already, we're growing potatoes that use 40% less chemical insecticide than would be possible using traditional techniques.

Quality traits — or "output" traits — help create value for consumers by enhancing the quality of the food and fiber produced by the plant. Likely future offerings include potatoes that will absorb less oil when fried, corn and soybeans with an increased protein content, tomatoes with a fresher flavor and strawberries that retain their natural sweetness.

Someday, seeds will become energy-efficient, environmentally friendly production facilities that can manufacture products which are today made from nonrenewable resources. A canola plant, for example, could serve as a factory to add beta carotene to canola oil to alleviate the nutritional deficiency that causes night blindness.

"The benefits of biotechnology are many and include providing resistance to crop pests to improve production and reduce chemical pesticide usage, thereby making major improvements in both food quality and nutrition."

NATIONAL BIOTECHNOLOGY POLICY-SALIENT FEATURES

The need for an integrated biotech policy with adequate attention to various sub sets of the sector such as health, agriculture, environment, industrial and other application areas is an essential pre-requisite for giving the boost to the progress of the Indian biotech sector. The National Science and Technology Policy of the Government and the Vision Statement on Biotechnology issued by the Department of Biotechnology (DBT) provide a framework and give strategic direction to different sectors to accelerate the pace of development of biotechnology in India. The policy has given a direction to efforts in the public and private sectors, the key being a quadrilateral agreement between academia, industry, various laboratories working in the field, and the state.

Salient Features of the Policy

- The need to augment the number of PhD programs in life sciences and biotechnology, as a strong pool of academic leaders is critical to sustained innovation. A National Task Force will be created to formulate model undergraduate and post graduate curricula, attract talent to life sciences and enable working conditions for scientists to undertake industry oriented research

- The need to scale up proven technologies such as diagnostics and vaccines. While the Indian industry is strong in product development and marketing for commercial benefits, biotech in India still lacks the infrastructure required for R&D in molecular modeling, protein engineering, drug designing and immunological studies. The DBT will act to facilitate a Single Window Clearance mechanism for the establishments of biotech plants and encourage private participation in infrastructure development
- India's strategy should aim at increasing value from R&D investment and IPR generation. India needs to provide active support through incubator funds and provision of various incentives, in addition to focusing on innovative capacity; the ability to create a continuous pipeline of products. Clear government policies for promotion of innovation and commercialisation of knowledge will propel growth of the biotechnology sector
- The need for government support, fiscal incentives and tax benefits are crucial to this sector, as biotechnology is the most research-intensive industry and companies invest 20-30 percent of their operating costs on R&D or technology outsourcing. Further, financial support for early phase product development and small/ medium size enterprises is the key to sustaining innovation
- Creation of Small Business Innovation Research Initiative (SBIRI) scheme through DBT to support small and medium size enterprises through grants and loans. The scheme will support pre-proof of concept, early stage innovative research and provide mentorship
- Establishment of biotechnology parks to provide a viable mechanism for licensing new technologies to upcoming biotech companies to start new ventures and achieve early stage value enhancement of technology with minimal financial inputs. Parks facilitate transfer of technology by serving as an impetus for entrepreneurship through partnership among innovators from academia, R&D institutions and industry.
- Need for a scientific, rigorous, transparent, efficient and consistent regulatory mechanism for biosafety evaluation; a single National Biotechnology Regulatory Authority be established and governed by an independent administrative structure.

India's food system is in deep crisis-50 million tonnes of food grains are rotting in the godowns while 300 million people go hungry and lakhs face starvation in Orissa, Rajasthan, Gujarat, and Chattisgarh. Even in spite of the crisis of hunger, the government has announced that it would dump two million tonnes of food grain into the sea to prevent spoilage and maintain high prices. The growing food stocks do not represent either over production or lack of storage capacity, but the lack of justice and the political will to defend the food rights of millions of Indians. Systematic dismantling of food security and farmers livelihoods The polarisation of wasted plenty while millions starve is a result of the deliberate and systematic dismantling of the national food security system and peoples food rights by the governments implementation of the World Bank/IMF Structural adjustment and the WTO's free trade policies.

* Cost of production are rising as globalisation of the input sector increases costs of seeds and agrochemical, privatisation increases the cost of power and irrigation, and subsidies related to agriculture are withdrawn, forcing farmers into suicide, land alienation and sale of body parts.
* Commodity prices are falling under the dual impact of the withdrawal of the government from procurement and failure to guarantee minimum support prices, as well as the dumping of artificially cheap and subsidised agricultural commodities through imports and the removal of quantitative restrictions. The prices of soyabean have collapsed from Rs.1300/qtl to 700/qtl. Mustard prices have dropped from Rs.1600/qtl to Rs.1000/qtl and coconut prices have collapsed from Rs.10 to Rs.2 per coconut.

RISING FOOD PRICES

The governments attempt to cut food subsidies by increasing food prices in fair price shops has led to declining off-take and declining consumption. Instead of lowering food prices for Indian consumers, the government now wants to hand over this food cheaply and increase their profit. While the price of wheat in the fair price shop is Rs.9.60, and Rs.4.60 for people below the poverty line, the government is to hand over wheat for exports at Rs.4.10. The government has also announced that to handle the crisis of growing stocks it will allow 100% FDI in storage and distribution.

SCIENCE AND TECHNOLOGY

India's food security has been based on the innovations, ingenuity and the wisdom of the small farmers. This sustainable indigenous scientific tradition was the inspiration for the founding of the Indian Science Congress. The Green Revolution ignored the needs of small farmers and the imperative of sustainable development and its only thrust was short-term high monoculture yield at the cost of the destruction of soil, water and biodiversity. At the beginning of the revolution the narrow goal may have been achieved, but after a span of two decades the farm productivity reached a plateau and is now showing a decline.

Without learning the lesson that non-sustainable models of farming cannot ensure food security, the government is blindly promoting genetic engineering on the false ground that this will increase food production and nutrition availability. The new agricultural policy has exclusive focus on transgenics. The Bt cotton trials were undertaken violating all concern for environmental laws and biosafety. Genetically engineered "golden rice" is being promoted as a cure for vitamin A deficiency and blindness even though it is inadequate in meeting the vitamin A needs. Meantime, easily available, cheap and rich sources of vitamin A in our indigenous fruits and vegetables are being destroyed through the use of agrochemical, such as Monsantoís Roundup and the spread of commercial monocultures. The introduction of herbicide resistant Roundup Ready crops will make the situation worse.

RESOURCE ALIENATION AND COMMODIFICATION OF COMMON RESOURCES

The incomplete agenda of land reforms is being reversed through introduction of land leasing and Corporate take over of land, as well as amendments in land acquisition laws. Indebtedness is also leading to large-scale land alienation and dispossession. The Ministry of Mines is seeking to remove all restrictions on the transfer of tribal and government lands in the schedule areas.

Further Land Acquisition Act of the 1894 (amended in 1984) is further proposed to be amended for faster and smoother acquisition of land by private companies without properly resettling and rehabilitating the affected population. The privatisation of water resources and power is also transferring water rights from farmers to corporations. New intellectual property rights regimes are threatening to turn seed into corporate property and farmers into bioserfs.

GLOBALISATION AND THE MARGINALISATION OF THE FARMER

The globalisation policies being blindly rushed into by government are part of a perverted paradigm that sees India only as a market for MNC products, and Indians only as consumers, and is blind to the millions of efficient small producers who have maintained India's food security. These policies of trade liberalisation are creating massive rural unemployment leading to poverty and a total collapse of purchasing power.

There has been a drastic decline in the number of cultivators while there has been a corresponding increase in the number of agricultural labourers. The marginal and displaced farmers and farm workers are thus being denied their fundamental right to food and livelihood guaranteed by the Indian Constitution.

The government has already announced the privatisation of grain procurement. Food distribution will therefore be driven by profits and not by peoples rights to food and livelihood.

OUR DEMANDS

Farmers are the best agricultural scientists especially for their own farm. Scientist make their best contributions when the support farmers and build the farmers knowledge. Instead anti farmer science/ technology is being thrust upon the farmers of India. Science should be developed for the people and not vice-versa. Further science developed over the years by the farmers need to be honoured.

The science that need to be developed should be aimed at more efficient resource use and not on farmers displacement. This resource depleting and livelihood-destroying paradigm of technology will and is resulting in mass unemployment, growing poverty and hunger. People's Science Congress therefore demands:

* Food and agriculture must be squarely excluded from the WTO agenda.
* The present Agriculture Policy, which is aimed at corporate takeover of Indian agriculture, reversing of land reforms and introduction of hazardous technologies like GMOs, must be scrapped, and must be replaced with a small and marginal farmer-centred agricultural policy.
* Government must take the responsibility to guarantee procurement at minimum support price. Adequate budgetary provisions must be made for intervention in unjust and unfair markets to ensure minimum support price to farmers.
* The government must stop unregulated imports, reintroduce QRs in farm commodities, and maintain them in others, in-act anti-dumping law and use both tariff and non-tariff protection for maintaining domestic food security and farmersí livelihood.
* Wherever farmers are subjected to payment of compound interest rates, that practice must be strongly opposed.
* The government must ensure social security systems for tribals, farmers and farm workers.
* Disaster relief entitlements have to be identified and managed directly by Gram Sabha. This must be statutorily guaranteed.

* The PDS system must be strengthened through transparency and public accountability as well as public participation in procurement and distribution of all commodities. The PDS system must reflect the food diversity and crop diversity of the country.
* The government should promote and strengthen cooperative and community storage systems in rural areas to ensure local food security.
* The government should institute systems implementation to ensure peopleís food rights through schemes such as food-for-work employment guarantee schemes, which also build infrastructure that enhances ecological security and food security and provides public services.
* The food and agriculture policy must be based on the efficiency and collective and cumulative wisdom of the Indian farmers.
* The untested and inefficient genetic engineering technologies should not be rushed to the market place, either as seed or as food.
* The violent and forcible collection of debts, which is forcing farmers into suicides and land alienation through auction and distress sale of land and property, must be immediately stopped.
* Seed business is farmersí business. Farmers' rights to seeds must be defended at all costs. There should be no monopoly in seed sector either through patents or plant breeders rights. Indian farmers cannot be criminalised for seed saving and seed exchange.

BIOTECHNOLOGY: INDIAN SCENARIO

India, the darling of the world as far as bio technology sector is concerned offers tremendous opportunities to companies to make investment in the sector in India.

Growth of all three areas of bio technology-medicinal, agricultural and industrial and conducive climate for the same make India as one of the ideal destination for world investment flow in India. Following are a few areas where opportunities exist for India:

VACCINES

India's huge and growing population makes it among the world's largest markets for vaccines of all types-India faces a growing demand for new-generation and 'combination' vaccines, such as DPT with Hepatitis B, Hepatitis A and injectable polio vaccine, besides several veterinary and poultry vaccines. Apart from conventional vaccines, the rDNA have further market potentials and offer great opportunities to companies in the economy.

MEDICINAL DISCOVERY

Opportunities exist for enhancing production facilities and economies of scale based on licensing, joint ventures, setting up of new production bases and establishing royalty sharing arrangements for all therapeutic and medicinal products approved for marketing in India, namely Insulin, Alpha, Interferon, Hepatitis B surface antigen based vaccine, Erythropoietin, Streptokinase, Chymotrypsin, and others.

AGRICULTURE SECTOR

Hybrid seeds, including genetically modified seeds such as Bt cotton represent new business opportunities based on yield improvement, and development of a production base in biopesticides and biofertilisers would facilitate India's entry into the growing organic or natural foods market.

The Genetically Modified crops like corn, cotton, millet, mustard and other nutritionally improved vegetables also provide good potential in the agriculture sector and also leads to improvement in farm produce and productivity per hectare.

MEDICINAL RESEARCH

New research and developments in the field make India as a hub of cutting edge technology for development of new products and medicines having ready and developed market for the same.

Indian pharmaceutical companies possess competitive skills in chemical synthesis and process engineering and extraction technologies, which they can leverage to develop new drugs and formulae. New investment into research and successful defending of patents by a number of Indian companies in ten world markets has opened new vistas of opportunities for Indian companies.

CLINICAL RESEARCH AND TRIALS

With clinical trials in India costing less than a fraction of what it costs in developed markets, clinical research organizations can seek research and trial projects in India from international companies, provided they are able to demonstrate best international practices and follow up procedures.

BIOINFORMATICS

Indian bioinformatics companies can play a significant role in critical areas such as data mining, lexican, mapping and DNA sequencing and extraction, molecule design simulation in world market for bilinformatics services. Complex algorithm writing and the use of computational capacities to study the 3D structures of proteins are the main skills required in this arena and India offers good investment opportunities for the same.

MODERN BIOTECHNOLOGY-SUSTAINABLE GROWTH

Modern biotechnology is purported to have a number of products for addressing certain food-security problems of developing countries. It offers the possibility of an agricultural system that is more reliant on biological processes rather than chemical applications.

The potential uses of modern biotechnology in agriculture include: increasing yields while reducing inputs of fertilizers, herbicides and insecticides; conferring drought or salt tolerance on crop plants; increasing shelf-life; reducing post harvest losses; increasing the nutrient content of produce; and delivering vaccines. The availability of such products could not only have an important role in reducing hunger and increasing food security, but also have the potential to address some of the health problems of the developing world.

Achieving the improvements in crop yields expected in developing countries can help to alleviate poverty: directly by increasing the household incomes of small farmers who adopt these technologies; and indirectly, through their positive impact as evidenced in the price slumps of herbicides and insecticides.

Indeed, some developing countries have identified priority areas such as tolerances to alkaline earth metals, drought and soil salinity, disease resistance, crop yields and nutritionally enhanced crops. The adoption of technologies designed to prolong shelf-life could be valuable in helping to reduce post harvest losses in regionally important crops. Prime candidates in terms of crops of choice for development are the so-called 'orphan crops', such as cassava, sweet potato, millet, sorghum and yam.

Currently, the many promises of modern biotechnology that could have an impact on food security have not been realized in most developing countries. In fact, the uptake of modern biotechnology has been remarkably low owing to the number of factors that underpin food security issues. In part, this could be because the first generation of commercially available crops using modern biotechnology were modified with single genes to impart resistance to pests, weed and insects, and not complex characteristics that would modify the growth of crops in harsh conditions. Secondly, the technologies are developed by companies in industrialized countries with little or no direct investment in, and which derive little economic benefit from, developing countries.

Although current commercial GM crops are not designed to address the specific issues of developing countries, their adoption has shown that they can be relevant in some developing countries-for example, the planting of herbicide-tolerant soybeans in Argentina and Bt cotton as a cash crop by resource-poor farmers in China and South Africa have resulted in significant gains for farmers.

On average, the Bt cotton farmers in China reduced pesticide spraying by 70%, producing a kilogram of cotton at 28% less cost than the non-Bt farmers. These benefits have had a significant impact on the agronomic, environmental, health and economic situations of approximately 5 million resource-poor farmers over eight provinces.

Several agro-economic studies have been commissioned since the introduction of seed derived from modern biotechnology in the USA. One report illustrates that the greatest yield increases were achieved with insect-resistant maize, while the greatest reduction of input costs was seen in herbicide-tolerant soybean.

There is lot more research required to be devoted to find specific solutions to food problems and use of biotechnology techniques in the area. Developing countries have more stake in these technologies and hence they would need to take lead in that.

2

Plant Genetic Engineering

Genetically modified plants are created by the process of genetic engineering, which allows scientists to move genetic material between organisms with the aim of changing their characteristics. All organisms are composed of cells that contain the DNA molecule. Molecules of DNA form units of genetic information, known as genes. Each organism has a genetic blueprint made up of DNA that determines the regulatory functions of its cells and thus the characteristics that make it unique. Prior to genetic engineering, the exchange of DNA material was possible only between individual organisms of the same species. With the advent of genetic engineering in 1972, scientists have been able to identify specific genes associated with desirable traits in one organism and transfer those genes across species boundaries into another organism.

A gene from bacteria, virus, or animal may be transferred into plants to produce genetically modified plants having changed characteristics. Thus, this method allows mixing of the genetic material among species that cannot otherwise breed naturally. The success of a genetically improved plant depends on the ability to grow single modified cells into whole plants. Some plants like potato and tomato grow easily from single cell or plant tissue. Others such as corn, soy bean, and wheat are more difficult to grow.

After years of research, plant specialists have been able to apply their knowledge of genetics to improve various crops such as corn, potato, and cotton. They have to be careful to ensure that the basic characteristics of these new plants are the same as the traditional ones, except for the addition of the improved traits. The world of biotechnology has always moved fast, and now it is moving even faster. More traits are emerging; more land than ever before is being planted with genetically modified varieties of an ever-expanding number of crops. Research efforts are being made to genetically modify most plants with a high economic value such as cereals, fruits, vegetables, and floriculture and horticulture species.

PLANT GENETICS

Knowledge of population genetics, quantitative genetics, probability theory and statistics is indispensable for understanding equilibria and shifts with regard to the genotypic composition of a population, its mean value and its variation. The subject of population genetics is the study of equilibria and shifts of allele and genotype frequencies in populations.

These equilibria and shifts are determined by five forces:

- Mode of reproduction of the considered crop

The mode of reproduction is of utmost importance with regard to the breeding of any particular crop and the maintenance of already available varieties. This applies both to the natural mode of reproduction of the crop and to enforced modes of reproduction, like those applied when producing a hybrid variety. In plant breeding theory, crops are therefore classified into the following categories: cross-fertilizing crops, self-fertilizing crops, crops with both cross- and self-fertilization and asexually reproducing crops. It is explained that even within a specific population, traits may differ with regard to their mode of reproduction.

- Selection
- Mutation
- Immigration of plants or pollen, *i.e.* immigration of alleles
- Random variation of allele frequencies

A population is a group of (potentially) interbreeding plants occurring in a certain area, or a group of plants originating from one or more common ancestors. The former situation refers to cross-fertilizing crops (in which case the term Mendelian population is sometimes used), while the latter group concerns, in particular, self-fertilizing crops. In the absence of immigration the population is said to be a closed population.

Examples of closed populations are:

- A group of plants belonging to a cross-fertilizing crop, grown in an isolated field, *e.g.* maize or rye (both pollinated by wind), or turnips or Brussels sprouts (both pollinated by insects)
- A collection of lines of a self-fertilizing crop, which have a common origin, *e.g.* a single-cross, a three-way cross, a backcross

The subject of quantitative genetics concerns the study of the effects of alleles and genotypes and of their interaction with environmental conditions. Population genetics is usually concerned with the probability distribution of genotypes within a population (genotypic composition), while quantitative genetics considers phenotypic values (and statistical parameters dealing with them, especially mean and variance) for the trait under investigation.

In fact population genetics and quantitative genetics are applications of probability theory in genetics. An important subject is, consequently, the derivation of probability distributions of genotypes and the derivation of expected genotypic values and of variances of genotypic values. Generally, statistical analyses comprise estimation of parameters and hypothesis testing. In quantitative genetics statistics is applied in a number of ways. It begins when considering the experimental design to be used for comparing entries in the breeding programme.

Considered across the entries constituting a population (plants, clones, lines, families) the expression of an observed trait is a random variable. If the expression is represented by a numerical value the variable is generally termed phenotypic value, represented by the symbol p.

Two genetic causes for variation in the expression of a trait are distinguished. Variation controlled by so-called major genes, *i.e.* alleles that exert a readily traceable effect on the expression of the trait, is called qualitative variation.

Variation controlled by so-called polygenes, *i.e.* alleles whose individual effects on a trait are small in comparison with the total variation, is called quantitative variation. In Note: it is elaborated that this classification does not perfectly coincide with the distinction between qualitative traits and quantitative traits.

The former paragraph suggests that the term *gene* and *allele* are synonyms. According to Rieger, Michaelis and Green a gene is a continuous region of DNA, corresponding to one (or more) transcription units and consisting of a particular sequence of nucleotides. Alternative forms of a particular gene are referred to as alleles. In this respect the two terms 'gene' and 'allele' are sometimes interchanged.

Thus the term 'gene frequency' is often used instead of the term 'allele frequency'. The term locus refers to the site, alongside a chromosome, of the gene/allele. Since the term 'gene' is often used as a synonym of the term 'locus', we have tried to avoid confusion by preferential use of the terms 'locus' and 'allele' (as a synonym of the word gene) where possible. In the case of qualitative variation, the phenotypic value p of an entry (plant, line, family) belonging to a genetically heterogeneous population is a discrete random variable. The phenotype is then exclusively (or to a largely traceable degree) a function f of the genotype, which is also a random variable G.

Thus, $p = f(G)$

It is often desired to deduce the genotype from the phenotype. This is possible with greater or lesser correctness, depending for example on the degree of dominance and sometimes also on the effect of the growing conditions on the phenotype. A knowledge of population genetics suffices for an insight into the dynamics of the genotypic composition of a population with regard to a trait with qualitative variation: application of quantitative genetics is then superfluous.

Note: All traits can show both qualitative and quantitative variation. Culm length in cereals, for instance, is controlled by dwarfing genes with major effects, as well as by polygenes. The commonly used distinction between qualitative traits and quantitative traits is thus, strictly speaking, incorrect. When exclusively considering qualitative variation, *e.g.* with regard to the traits in pea (*Pisum sativum*) studied by Mendel, this book describes the involved trait as a trait showing qualitative variation.

On the other hand, with regard to traits where quantitative variation dominates – and which are consequently mainly discussed in terms of this variation – one should realise that they can also show qualitative variation.

In this sense the following economically important traits are often considered to be 'quantitative characters':

- Biomass
- Yield with regard to a desired plant product
- Content of a desired chemical compound (oil, starch, sugar, protein, lysine) or an undesired compound
- Resistance, including components of partial resistance, against biotic or abiotic stress factors
- Plant height

In the case of quantitative variation p results from the interaction of a complex genotype, *i.e.* several to many loci are involved, and the specific growing conditions are important. In this book, by complex genotype we mean the sum of the genetic constitutions of all loci affecting the expression of the considered trait.

These loci may comprise loci with minor genes (or polygenes), as well as loci with major genes, as well as loci with both. With regard to a trait showing quantitative variation, it is impossible to classify individual plants, belonging to a genetically heterogeneous population, according to their genotypes.

This is due to the number of loci involved and the complicating effect on p of (some) variation in the quality of the growing conditions. It is, thus, impossible to determine the number of plants representing a specified complex genotype. (With regard to the expression of qualitative variation this may be possible!). Knowledge of both population genetics and quantitative genetics is therefore required for an insight into the inheritance of a trait with quantitative variation.

The phenotypic value for a quantitative trait is a continuous random variable and so one may write:

$$p = f(G,e)$$

Thus the phenotypic value is a function f of both the complex genotype (represented by G) and the quality of the growing conditions (say environment, represented by e).

Even in the case of a genetically homogeneous group of plants (a clone, a pure line, a single-cross hybrid) p is a continuous random variable.

The genotype is a constant and one should then write:

$$p = f(G,e)$$

Regularly in this book, simplifying assumptions will be made when developing quantitative genetic theory.

Especially the following assumptions will often be made:

- Absence of linkage of the loci controlling the studied trait(s)
- Absence of epistatic effects of the loci involved in complex genotypes.

These assumptions will now be considered.

Absence of Linkage

The assumption of absence of linkage for the loci controlling the trait of interest, *i.e.* the assumption of independent segregation, may be questionable in specific cases, but as a generalisation it can be justified by the following reasoning.

Suppose that each of the n chromosomes in the genome contains M loci affecting the considered trait. This implies presence of n groups of,

$$\binom{M}{2} -$$

pairs of loci consisting of loci which are more strongly or more weakly linked. The proportion of pairs consisting of linked loci among all pairs of loci amounts then to,

$$\frac{n\binom{M}{2}}{\binom{nM}{2}} = \frac{n.M!}{2!(M-2)!} \times \frac{2!(nM-2)!}{(nM)!} = \frac{M-1}{nM-1} = \frac{1-\dfrac{1}{M}}{n-\dfrac{1}{M}}$$

For $M = 1$ this proportion is 0; for $M = 2$ it amounts to 0.077 for rye (*Secale cereale*, with $n = 7$) and to 0.024 for wheat (*Triticum aestivum*, with $n = 21$); for $M = 3$ it amounts to 0.100 for rye and to 0.032 for wheat. For $M \to \infty$ the proportion is 1 n; *i.e.* 0.142 for rye and 0.048 for wheat.

One may suppose that loci located on the same chromosome, but on different sides of the centromere, behave as unlinked loci. If each of the n chromosomes contains,

$$m\left(=\frac{1}{2}M\right)$$

relevant loci on each of the two arms then there are $2n$ groups of,

$$\binom{m}{2}-$$

pairs consisting of linked loci. Thus considered, the proportion of pairs consisting of linked loci amounts to,

$$\frac{2n\binom{m}{2}}{\binom{2nm}{2}}=\frac{2n.m!}{2!(m-2)!}\times\frac{2!(2nm-2)!}{(2nm)!}=\frac{1-\frac{1}{m}}{2n-\frac{1}{m}}$$

For $m = 1$ this proportion is 0; for $m = 2$ it amounts to 0.037 for rye and to 0.012 for wheat; for $m = 3$ it amounts to 0.049 for rye and to 0.016 for wheat. For $m \to \infty$ the proportion is,

$$\frac{1}{2n}.e.\ 0.071 \text{ for rye and } 0.024 \text{ for wheat.}$$

For the case of an even distribution across all chromosomes of the polygenic loci affecting the considered trait it is concluded that the proportion of pairs of linked loci tends to be low. (In an autotetraploid crop the chromosome number amounts to $2n = 4x$. The reader might like to consider what this implies for the above expressions.)

Absence of epistasis

Absence of epistasis is another assumption that will be made regularly in this book. It implies additivity of the effects of the single-locus genotypes for the loci affecting the level of expression for the considered trait. The genotypic value of some complex genotype consists then of the sum of the genotypic value of the complex genotype with regard to all non-segregating loci, here represented by *m*, as well as the sum of the contributions due to the genotypes for each of the *K* segregating polygenic loci *B*1-*b*1..., *BK*-*bK*. Thus,

$$GB_1-b_1....,B_k-b_k=m+G'B_1-b_1+...+G'B_k-b_k$$

where G_ is defined as the contribution to the genotypic value, relative to the population mean genotypic value, due to the genotype for the considered locus.

The assumption implies the absence of inter-locus interaction, *i.e.* the absence of epistasis (in other words: absence of non-allelic interaction).

It says that the effect of some genotype for some locus *Bi-bi* in comparison to another genotype for this same locus does not depend at all on the complex genotype determined by all other relevant loci. In this book, in order to clarify or substantiate the main text, theoretical examples and results of actual experiments are presented. Notes provide short additional information and appendices longer, more complex supplementary information or mathematical derivations.

UNDERSTANDING GENE TESTING

Genes- the chemical messages of heredity- constitute a blueprint of our possibilities and limitations. The legacy of generations of ancestors, our genes carry the key to our similarities and our uniqueness.

When genes are working properly, our bodies develop and function smoothly. But should a single gene-even a tiny segment of a single gene- go awry, the effect can be dramatic: deformities and disease, even death.

In the past 20 years, amazing new techniques have allowed scientists to learn a great deal about how genes work and how genes are linked to disease. Increasingly, researchers are able to identify mutations, changes within genes that can lead to specific disorders. Tests for gene mutations make it possible not only to detect diseases already in progress but also, in certain situations, to foresee diseases yet to come.

This new ability raises both high hopes and grave concerns. On the one hand, predictive gene testing holds out the possibility of saving thousands of lives through prevention or early detection. On the other, the implications of test results are enormous, not only for the individual but also for relatives who share this genetic legacy, and for society as a whole.

This booklet presents key concepts and issues relevant to gene testing and answers questions that are frequently asked about the techniques, potential risks, and possible benefits of attempts to link genetic markers with disease. Words that appear underlined in the pages that follow are defined in the Glossary. Clicking on these words will take you to the glossary. From there press the browser's back button to return.

What are Genes?

Genes are working subunits of DNA. DNA is a vast chemical information database that carries the complete set of instructions for making all the proteins a cell will ever need. Each gene contains a particular set of instructions, usually coding for a particular protein.

DNA exists as two long, paired strands spiraled into the famous double helix. Each strand is made up of millions of chemical building blocks called bases. While there are only four different chemical bases in DNA (adenine, thymine, cytosine, and guanine), the order in which the bases occur determines the information available, much as specific letters of the alphabet combine to form words and sentences.

DNA resides in the core, or nucleus, of each of the body's trillions of cells. Every human cell (with the exception of mature red blood cells, which have no nucleus) contains the same DNA. Each cell has 46 molecules of double-stranded DNA. Each molecule is made up of 50 to 250 million bases housed in a chromosome.

The DNA in each chromosome constitutes many genes (as well as vast stretches of noncoding DNA, the function of which is unknown). A gene is any given segment along the DNA that encodes instructions that allow a cell to produce a specific product- typically, a protein such as an enzyme- that initiates one specific action. There are between 50,000 and 100,000 genes, and every gene is made up of thousands, even hundreds of thousands, of chemical bases.

How do Genes Work?

Although each cell contains a full complement of DNA, cells use genes selectively. Some genes enable cells to make proteins needed for basic functions; dubbed housekeeping genes, they are active in many types of cells. Other genes, however, are inactive most of the time. Some genes play a role in early development of the embryo and are then shut down forever. Many genes encode proteins that are unique to a particular kind of cell and that give the cell its character- making a brain cell, say, different from a bone cell. A normal cell activates just the genes it needs at the moment and actively suppresses the rest.

RNA AND DNA REVEALED: NEW ROLES, NEW RULES

For many years, when scientists thought about heredity, DNA was the first thing to come to mind. It's true that DNA is the basic ingredient of our genes and, as such, it often steals the limelight from RNA, the other form of genetic material inside our cells.

But, while they are both types of genetic material, RNA and DNA are rather different. The chemical units of RNA are like those of DNA, except that RNA has the nucleotide uracil (U) instead of thymine (T). Unlike double-stranded DNA, RNA usually comes as only a single strand. And the nucleotides in RNA contain ribose sugar molecules in place of deoxyribose. RNA is quite flexible—unlike DNA, which is a rigid, spiral-staircase molecule that is very stable. RNA can twist itself into a variety of complicated, three-dimensional shapes. RNA is also unstable in that cells constantly break it down and must continually make it fresh, while DNA is not broken down often. RNA's instability lets cells change their patterns of protein synthesis very quickly in response to what's going on around them.

Many textbooks still portray RNA as a passive molecule, simply a "middle step" in the cell's gene-reading activities. But that view is no longer accurate. Each year, researchers unlock new secrets about RNA. These discoveries reveal that it is truly a remarkable molecule and a multi-talented actor in heredity.

Today, many scientists believe that RNA evolved on the Earth long before DNA did. Researchers hypothesize—obviously, no one was around to write this down—that RNA was a major participant in the chemical reactions that ultimately spawned the first signs of life on the planet.

RNA World

At least two basic requirements exist for making a cell: the ability to hook molecules together and break them apart, and the ability to replicate, or copy itself, from existing information. RNA probably helped to form the first cell. The first organic molecules, meaning molecules containing carbon, most likely arose out of random collisions of gases in the Earth's primitive atmosphere, energy from the Sun, and heat from naturally occurring radioactivity. Some scientists think that in this primitive world, RNA was a critical molecule because of its ability to lead a double life: to store information and to conduct chemical reactions. In other words, in this world, RNA served the functions of both DNA and proteins.

What does any of this have to do with human health? Plenty, it turns out.

Today's researchers are harnessing some of RNA's flexibility and power. For example, through a strategy he calls directed evolution, molecular engineer Ronald R. Breaker of Yale University is developing ways to create entirely new forms of RNA and DNA that both work as enzymes.

Recently, Breaker and others have also uncovered a hidden world of RNAs that play a major role in controlling gene activity, a job once thought to be performed exclusively by proteins. These RNAs, which the scientists named riboswitches, are found in a wide variety of bacteria and other organisms.

This discovery has led Breaker to speculate that new kinds of antibiotic medicines could be developed to target bacterial riboswitches.

Molecular Editor

Scientists are learning of another way to customize proteins: by RNA editing. Although DNA sequences spell out instructions for producing RNA and proteins, these instructions aren't always followed precisely. Editing a gene's mRNA, even by a single chemical letter, can radically change the resulting protein's function. Nature likely evolved the RNA editing function as a way to get more proteins out of the same number of genes. For example, researchers have found that the mRNAs for certain proteins important for the proper functioning of the nervous system are particularly prone to editing. It may be that RNA editing gives certain brain cells the capacity to react quickly to a changing environment.

Which molecules serve as the editor and how does this happen? Brenda Bass of the University of Utah School of Medicine in Salt Lake City studies one particular class of editors called adenosine deaminases. These enzymes "retype" RNA letters at various places within an mRNA transcript.

They do their job by searching for characteristic RNA shapes. Telltale twists and bends in folded RNA molecules signal these enzymes to change the RNA sequence, which in turn changes the protein that gets made.

Bass' experiments show that RNA editing occurs in a variety of organisms, including people. Another interesting aspect of editing is that certain disease-causing microorganisms, such as some forms of parasites, use RNA editing to gain a survival edge when living in a human host. Understanding the details of this process is an important area of medical research.

Small But Powerful

Recently, molecules called microRNAs have been found in organisms as diverse as plants, worms, and people. The molecules are truly "micro," consisting of only a few dozen nucleotides, compared to typical human mRNAs that are a few thousand nucleotides long.

What's particularly interesting about microRNAs is that many of them arise from DNA that used to be considered merely filler material.

How do these small but important RNA molecules do their work? They start out much bigger but get trimmed by cellular enzymes, including one aptly named Dicer. Like tiny pieces of Velcro®, microRNAs stick to certain mRNA molecules and stop them from passing on their protein-making instructions.

First discovered in a roundworm model system some microRNAs help determine the organism's body plan. In their absence, very bad things can happen. For example, worms engineered to lack a microRNA called let-7 develop so abnormally that they often rupture and practically break in half as the worm grows.

Perhaps it is not surprising that since microRNAs help specify the timing of an organism's developmental plan, the appearance of the microRNAs themselves is carefully timed inside a developing organism. Biologists, including Amy Pasquinelli of the University of California, San Diego, are currently figuring out how microRNAs are made and cut to size, as well as how they are produced at the proper time during development.

MicroRNA molecules also have been linked to cancer. For example, Gregory Hannon of the Cold Spring Harbor Laboratory on Long Island, New York, found that certain microRNAs are associated with the severity of the blood cancer B-cell lymphoma in mice. Since the discovery of microRNAs in the first years of the 21st century, scientists have identified hundreds of them that likely exist as part of a large family with similar nucleotide sequences. New roles for these molecules are still being found.

RNA INTERFERENCE (RNAI)

Healthy Interference

RNA controls genes in a way that was only discovered recently: a process called RNA interference, or RNAi. Although scientists identified RNAi less than 10 years ago, they now know that organisms have been using this trick for millions of years.

Researchers believe that RNAi arose as a way to reduce the production of a gene's encoded protein for purposes of fine-tuning growth or self-defence. When viruses infect cells, for example, they command their host to produce specialized RNAs that allow the virus to survive and make copies of itself. Researchers believe that RNAi eliminates unwanted viral RNA, and some speculate that it may even play a role in human immunity.

Oddly enough, scientists discovered RNAi from a failed experiment! Researchers investigating genes involved in plant growth noticed something strange: When they tried to turn petunia flowers purple by adding an extra "purple" gene, the flowers bloomed white instead.

This result fascinated researchers, who could not understand how adding genetic material could somehow get rid of an inherited trait. The mystery remained unsolved until, a few years later, two geneticists studying development saw a similar thing happening in lab animals.

The researchers, Andrew Z. Fire, then of the Carnegie Institution of Washington in Baltimore and now at Stanford University, and Craig Mello of the University of Massachusetts Medical School in Worcester, were trying to block the expression of genes that affect cell growth and tissue formation in roundworms, using a molecular tool called antisense RNA.

To their surprise, Mello and Fire found that their antisense RNA tool wasn't doing much at all. Rather, they determined, a doublestranded contaminant produced during the synthesis of the single-stranded antisense

RNA interfered with gene expression. Mello and Fire named the process RNAi, and in 2006 were awarded the Nobel Prize in physiology or medicine for their discovery.

Further experiments revealed that the doublestranded RNA gets chopped up inside the cell into much smaller pieces that stick to mRNA and block its action, much like the microRNA pieces of Velcro discussed above.

Today, scientists are taking a cue from nature and using RNAi to explore biology. They have learned, for example, that the process is not limited to worms and plants, but operates in humans too.

Medical researchers are currently testing new types of RNAi-based drugs for treating conditions such as macular degeneration, the leading cause of blindness, and various infections, including those caused by HIV and herpes virus.

Dynamic DNA

A good part of who we are is "written in our genes," inherited from Mom and Dad. Many traits, like red or brown hair, body shape, and even some personality quirks, are passed on from parent to offspring.

But genes are not the whole story. Where we live, how much we exercise, what we eat: These and many other environmental factors can all affect how our genes get expressed.

You know that changes in DNA and RNA can produce changes in proteins. But additional control happens at the level of DNA, even though these changes do not alter DNA directly. Inherited factors that do not change the DNA sequence of nucleotides are called epigenetic changes, and they too help make each of us unique.

Epigenetic means, literally, "upon" or "over" genetics. It describes a type of chemical reaction that can alter the physical properties of DNA without changing its sequence. These changes make genes either more or less likely to be expressed.

Currently, scientists are following an intriguing course of discovery to identify epigenetic factors that, along with diet and other environmental influences, affect who we are and what type of illnesses we might get.

Secret Code

DNA is spooled up compactly inside cells in an arrangement called chromatin. This packaging is critical for DNA to do its work. Chromatin consists of long strings of DNA spooled around a compact assembly of proteins called histones.

One of the key functions of chromatin is to control access to genes, since not all genes are turned on at the same time. Improper expression of growth-promoting genes, for example, can lead to cancer, birth defects, or other health concerns.

Many years after the structure of DNA was determined, researchers used a powerful device known as an electron microscope to take pictures of chromatin fibres. Upon viewing chromatin up close, the researchers described it as "beads on a string," an image still used today. The beads were the histone balls, and the string was DNA wrapped around the histones and connecting one bead to the next.

Decades of study eventually revealed that histones have special chemical tags that act like switches to control access to the DNA. Flipping these switches, called epigenetic markings, unwinds the spooled DNA so the genes can be transcribed.

The observation that a cell's gene-reading machinery tracks epigenetic markings led C. David Allis, who was then at the University of Virginia Health Sciences Centre in Charlottesville and now works at the Rockefeller University in New York City, to coin a new phrase, the "histone code." He and others believe that the histone code plays a major role in determining which proteins get made in a cell.

Flaws in the histone code have been associated with several types of cancer, and researchers are actively pursuing the development of medicines to correct such errors.

Genetics and You: The Genetics of Anticipation

Occasionally, unusual factors influence whether or not a child will be born with a genetic disease.

An example is the molecular error that causes Fragile X syndrome, a rare condition associated with mental retardation. The mutation leading to a fragile X chromosome is not a typical DNA typing mistake, in which nucleotides are switched around or dropped, or one of them is switched for another nucleotide. Instead, it is a kind of stutter by the DNA polymerase enzyme that copies DNA. This stutter creates a string of repeats of a DNA sequence that is composed of just three nucleotides, CGG.

The number of triplet repeats seems to increase as the chromosome is passed down through several generations. Thus, the grandsons of a man with a fragile X chromosome, who is not himself affected, have a 40 percent risk of retardation if they inherit the repeat-containing chromosome. The risk for great-grandsons is even higher: 50 percent.

Intrigued by the evidence that triplet repeats can cause genetic disease, scientists have searched for other examples of disorders associated with the DNA expansions. To date, more than a dozen such disorders have been found, and all of them affect the nervous system.

Analysis of the rare families in which such diseases are common has revealed that expansion of the triplet repeats is linked to something called genetic anticipation, when a disease's symptoms appear earlier and more severely in each successive generation.

BIOLOGICAL CELL

It only takes one biological cell to create an organism. In fact, there are countless species of single celled organisms, and indeed multi-cellular organisms like ourselves.

A single cell is able to keep itself functional by owning a series of '*miniature machines*' known as *organelles*. The following list looks at some of these organelles and other characteristics typical of a fully functioning cell.

- *Mitochondrion:* An important cell organelle involved in respiration
- *Cytoplasm:* A fluid surrounding the contents of a cell and forms a vacuole
- *Golgi Apparatus:* The processing area for the creation of a glycoprotein
- *Endoplasmic Reticulum:* An important organelle heavily involved in protein synthesis.
- *Vesicles: Packages* of substances that are to be used in the cell or secreted by it.
- *Nucleus*: The "brain" of a cell containing genetic information that determines every natural process within an organism.
- *Cell Membrane:* Also known as a plasma membrane, this outer layer of a cell assists in the movement of molecules in and out the cell plays both a structural and protective role
- *Lysosomes:* Membranous sacs that contain digestive enzymes

Cell Wall

A structure that characteristically is found in plants and prokaryotes and not animals that plays a structural and protective role.

Cell Specialisation

Cells can become specialised to perform a particular function within an organism, usually as part of a larger tissue consisting of many of the same cells working in tandem:

- *Nerve cells* to operate as part of the nervous system to send messages back and forth via the brain at the centre of the nerve system.
- Skin cells for waterproof protection and protection against pathogens in the open air environment.

- *Xylem* tubes to transport water around plants and to provide structural support for the plant as a whole.

Cells combine their efforts in these tissue types to perform a common cause. The task of the specialised cell will determine in what way it is going to be specialised, because different cells are suited to different purposes, as illustrated in the above list:

- Muscle cells are long and smooth in structure and their elastic nature allows these cells to perform flexible movements, just as they do in our own body's.
- Some *white blood cells* contain powerful digestive enzymes to eliminate pathogens by breaking them down to the molecular level.
- Cells at the back of the eye are sensitive to light stimuli, and thus can *interpret* differences in light intensity which can in turn be interpreted by our nervous system and brain.

Many of these cells contain organelles, though after some cells are specialised, they do not possess particular characteristics as they do not require them to be there. *i.e.* efficiency is the key, no resources are wasted and the resources available are put to their idyllic optimum.

The Cell Membrane

The *cell membrane,* otherwise known as the plasma membrane is a semi-permeable structure consisting mainly of *phospholipid* (fat) molecules and proteins. They are structured in a *fluid mosaic model,* where a double layer of phospholipid molecules provide a barrier accompanied by proteins. It is present round the circumference of a cell to acts as a barrier, keeping foreign entities out the cell and its contents (like cytoplasm) firmly inside the cell. The plasma membrane allows only selected materials to pass in and out of a cell, and is thus known as a selectively permeable membrane.

Cell Transport

There are three methods in which ions are transported through the cell membrane into the cell,

- *Active Transport*: Active transport is the transport of molecules with the active assistance of a carrier that can transport the material against a natural *concentration gradient.*
- *Passive Transport (Diffusion):* The movement of molecules from areas of high concentration (*i.e.* outside a cell) to areas of low concentration (*i.e.* within a cell) via a carrier. This process does not require energy.
- *Simple Diffusion:* The movement of molecules from areas of high concentration to areas of low concentration in a free state. *Osmosis* of

water involves this type of diffusion through a selectively permeable membrane (*i.e.* plasma membrane)

The Breakdown of Materials in a Cell

In cells, sometimes it is required to breakdown more complex molecules into more simple molecules, which can then be 're-built' into what is needed by the body with these new raw materials. *'Pinocytosis'* where to contents of a structure (such as bacteria) are *drank*, essentially by breaking down molecules into a drinkable form. *'Phagocytosis'* where contents are 'eaten'.

Absorption and Secretion

Absorption is the uptake of materials from a cells' external environment. Secretion is the ejection of material.

BIOLOGICAL ENERGY-ADP AND ATP

ATP stands for Adenosine Tri-Phosphate, and is the energy used by an organism in its daily operations. It consists of an *adenosine* molecule and three inorganic *phosphates*. After a simple reaction breaking down ATP to *ADP*, the energy released from the breaking of a molecular bond is the energy we use to keep ourselves alive.

ATP to ADP Energy Release

This is done by a simple process, in which one of the phosphate molecules is broken off, therefore reducing the ATP from 3 phosphates to 2, forming ADP (Adenosine Diphosphate after removing one of the phosphates {Pi}). This is commonly wrote as ADP + Pi.

When the bond connecting the phosphate is broken, *energy* is released. While ATP is constantly being used up by the body in its biological processes, the energy supply can be bolstered by new sources of glucose being made available via eating food which is then broken down by the digestive system to smaller particles that can be utilised by the body. On top of this, ADP is built back up into ATP so that it can be used again in its more energetic state. Although this conversion requires energy, the process produces a net gain in energy, meaning that more energy is available by re-using ADP+Pi back into ATP.

Glucose and ATP

Many ATP are needed every second by a cell, so ATP is created inside them due to the demand, and the fact that organisms like ourselves are made up of millions of cells.

Glucose, a sugar that is delivered via the bloodstream, is the product of the food you eat, and this is the molecule that is used to create ATP. Sweet foods provide a rich source of readily available glucose while other foods provide the materials needed to create glucose. This glucose is broken down in a series of *enzyme* controlled steps that allow the release of energy to be used by the organism. This process is called respiration.

Respiration and the Creation of ATP

ATP is created via respiration in both animals and plants. The difference with plants is the fact they attain their food from elsewhere. In essence, materials are harnessed to create ATP for biological processes. The energy can be created via cell respiration.

The process of respiration occurs in 3 steps (when oxygen is present:

- Glycolysis
- The Kreb's Cycle
- The Cytochrome System

CELL RESPIRATION

As mentioned in the previous page on ATP, the process of *respiration* is split into 3 distinct areas that occur at different parts of the cell. Respiration involves the *oxidation* of foodstuff (*i.e.* glucose) in order to create ATP. Respiration can occur with or without oxygen, *aerobic* and *anaerobic* respiration respectively.

Glycolysis

Glycolysis occurs in the *cytoplasm* of a cell where a 6 carbon glucose molecule (the broken down food that you ate earlier) is broken down by enzymes into a 3 carbon *pyruvic acid*. The execution of this process requires 2 ATP, and produces a net gain of 2 ATP. The enzymes involved remove hydrogen from the glucose (oxidation) where they take these hydrogen atoms to the cytochrome system, explained soon. In anaerobic respiration, this is where the process ends, glucose is split into 2 molecules of pyruvic acid. When oxygen is present, pyruvic is broken down into other carbon compounds in the Kreb's Cycle. When it is not present, the pyruvic acid is broken down into lactic acid (or carbon dioxide and ethanol).

The Kreb's Cycle

When oxygen is present, respiration can harness more ATP from a single unit of glucose. The *pyruvic acid* from the *glycolysis* stage diffuses into a cell organelle called a mitochondrion (pl. *mitochondria*). These mitochondria are sausage shaped structures that host a large surface area for the respiration to occur on. The pyruvic acid is then subject to more enzymes which break it down into a 2 carbon compound, as seen below.

The diagram illustrates the Kreb's cycle, consisting of three main actions:

- The carbon element is in an infinite cycle where the 2 carbon compound derived from pyruvic acid binds with the 4 carbon compound that is always present in the cycle.
- CO_2 is released, where the oxygen that is present in *aerobic respiration* combines with carbon from the carbon compounds which is released as CO_2. Hence the need for animals to breath out and expel this CO_2.
- Enzymes oxidize the carbon compounds and transport the hydrogen atoms to the cytochrome system.

The Cytochrome System

The cytochrome system, also known as the hydrogen carrier system (or the *electron transport system*) are where the reduced hydrogen carriers transport hydrogen atoms from the glycolysis and Kreb's cycle stages. The cytochrome system is found in the many *cristae* of mitochondria, which are tiny stalked particles found on its outer layer.

The system contains many 'hydrogen acceptors' which hydrogen can be added to. By following the path of a hydrogen atom, we can see how the cytochrome system works:

- Some *coenzymes* from earlier stages are transferred to the next coenzymes.
- B is then oxidised, therefore the coenzyme releases the hydrogen and energy is made available.
- The released hydrogen atom binds with 2 oxygen atoms (oxygen is available in aerobic respiration) which produces water, a by-product of respiration.

The diagram illustrates this flow of hydrogen within the cytochrome system and how energy is made available by the flow of these atoms. The green circles illustrate where energy is made available via oxidation. Overall their is a gain of 38 ATP from one molecule of glucose in aerobic respiration. The food that we eat provides glucose required in respiration. In plants, energy is also acquired via respiration, but the mechanism of delivering glucose to the respiration process is a little different.

BIOLOGICAL VIRUSES

The prime directive of all organisms is to reproduce and survive, which is also the case for viruses, which in most cases are considered a nuisance to humans.

Viruses

Viruses possess both living and non-living characteristics. The unique characteristic that differentiates viruses from other organisms is the fact that they require other organisms to host themselves in order to survive, hence they are deemed *obligate parasites*.

Viruses can be spread in the following exemplar ways:

- *Airborne*: Viruses that infect their hosts from the open air
- *Blood Borne*: Transmission of the virus between organisms when infected blood enters an organisms circulatory system
- *Contamination*: Caused from the consumption of materials by organisms such as water and food which have viruses within

Therefore viruses have many means of getting transmitted from one organism to another.

Cell Assimilation by a Virus

Viruses are tiny micro-organisms, and due to their size and simplicity, they are unable to replicate independently. Therefore, when a virus is situated in a host, it requires the means to reproduce before it dies out without producing more viruses.

This is done by altering the genetic make up of a cell to start coding for materials required to make more viruses. By altering the cell instructions, more viruses can be produced which in turn, can affect more cells and continue their existence as a species. The following is a step by step guide of how an example bacteriophage (a virus that infects bacteria) takes control of its host cell and reproduces itself.

- The virus approaches the bacteria and attaches itself to the cell membrane
- The tail gives the virus the means to thrust its genetic information into the bacteria
- Nucleotides from the host are 'stolen' in order for the virus to create copies of itself
- The viral DNA alters the genetic coding of the host cell to create protein coats for the newly create viral DNA strands
- The viral DNA enters its DNA coat
- The cell is swollen with many copies of the original virus and bursts, allowing the viruses to attach themselves to other nearby cells
- The process begins all over again with many more viruses attacking the hosts' cells

Without a means of defence, the host that is under attack from the virus would soon die.

Biological Cell Defence

Organisms must find a means of defence against antigens such a viruses described on the previous page. If this was not the case, bacteria, fungi and viruses would replicate out of control inside other organisms which would most likely already be extinct.

Therefore organisms employ many types of defence to stop this happening. Means of defence can be categorised into first and second lines of defence, with the first line usually having direct contact with the external environment.

First Lines of Defence

- *Skin* is an excellent line of defence because it provides an almost impenetrable biological barrier protecting the internal environment.
- *Lysozyme* is an enzyme found in tears and saliva that has powerful digestive capabilities, and can break down foreign agents to a harmless status before they enter the body.
- The clotting of blood near open wounds prevents an open space for antigens to easily enter the organism by coagulating the blood.
- *Mucus* and *cilia* found in the nose and throat can catch foreign agents entering these open cavities then sweep them outside via coughing, sneezing and vomiting.
- The *cell wall* of plants consists of fibrous proteins which provide a barrier to potential parasites (antigens).

If these first lines of defence fail, then there are further defences found within the body to ensure that the foreign agent is eliminated.

Second Lines of Defence

Second lines of defence deal with antigens that have bypassed the first lines of defence and still remain a threat to the infected organism. *Interferons* are a family of proteins that are released by a cell that is under attack by an antigen. These interferons attach themselves to *receptors* on the plasma membrane of other cells, effectively instructing it of the previous cells' situation.

This tells these neighbouring cells that an antigen is nearby and instructs them to begin coding for antiviral proteins, which upon action, defend the cell by shutting it down. In light of this, any invading antigen will not be able to replicated its DNA (or mRNA) and *protein coat* inside the cell, effectively preventing the spread of it in the organism. These antiviral proteins provide the organism with protection against a wide range of viruses.

This action brought about by interferon is a defensive measure, while *white blood cells* in the second line of defence in animals can provide a means of attacking these antigens. One method of attacking antigens is by a method called *phagocytosis*, where the contents of the antigen are broken down by molecules called phagocytes. These phagocytes contain digestive enzymes in their lysosomes (an organelle in phagocytes) such as lysozyme. White blood cells such as a *neutrophil* or a monocyte are capable of undergoing phagocytosis, which is illustrated below.

- The bacterium inside the cell gives out chemical messages that are picked up by the phagocyte.
- The bacteria targets the cell as a possible host and moves towards it.
- The cell is prepared for this and the bacterium becomes trapped in a vacuole that forms around it.
- The bacterium is a sitting duck that is harmless at present.
- The lysosomes detect the bacterium and the digestive enzymes inside them begin to break the bacterium down.
- The remnants of the lysosome and bacterium materials are absorbed into the cytoplasm.

The above illustrates one method of ridding an organism of an internal threat caused by an antigen. This is a non-specific response to an antigen.

Passive and Active Types of Immunity

The previous page investigated the role of white blood cells in phagocytosis. *White blood cells* are also responsible for *antibody* formation. Certain antibodies are synthesised in response to the presence of certain *antigens*

Specific Immune Responses

Lymphocytes are a type of white blood cell capable of producing a *specific immune response* to unique antigens. Some of these lymphocytes are capable of entrapping antigens on their surface. When lymphocytes catch these antigens they can then begin to code for unique antibodies, structures that are capable of catching these antigens. The lymphocytes code for a particular antibody on response to a particular antigen. The antibody that is formed will be capable of catching free antigens therefore neutralising the threat as seen below.

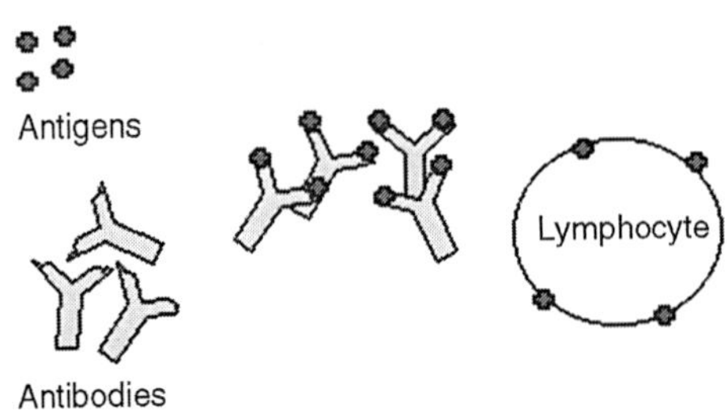

B lymphocytes (B Cells) produce free moving antibodies as above while *T lymphocytes* (T Cells) produce antibodies on their surface.

Types of Immunity

When attacked an organism has several means in which it can prepare to defend itself in event of attack.

- *Active Immunity*: *Vaccines* are used for health purposes to expose our bodies to a particular antigen. These antigens are usually killed or severely weakened to decrease their potency. After destroying these pathogens, the body stores some T cells as memory cells, due to the fact they code for a particular antigen and can be when needed. This memory in T cells can be a means of artificially acquiring immunity while a genuine attack by a pathogen is a naturally acquired type of immunity.
- *Passive Immunity*: This is where immunity to particular antigens as a result of genetic traits passed on from parents rendering the offspring immune to a particular pathogenic threat.

Plant transformation

Genetic engineering of plants is much easier than that of animals.

There are several reasons for this:

- There is a natural transformation system for plants (the bacterium *Agrobacterium tumefaciens*),
- Plant tissue can redifferentiate (a transformed piece of leaf may be regenerated to a whole plant), and
- Plant transformation and regeneration are relatively easy for a variety of plants.

The soil bacterium Agrobacterium tumefaciens ("tumefaciens" meaning tumor-making) can infect wounded plant tissue, transferring a large plasmid, the Ti plasmid, to the plant cell. Part of the Ti (tumor-inducing) plasmid apparently randomly integrates into the chromosome of the plant.

The integrated part of the plasmid contains genes for the synthesis of:

- Food for the bacterium, and
- *Plant hormones*: Genes from the Ti plasmid that are integrated in the plant chromosome are expressed at high levels in the plant.

Overproduction of the plant hormones leads to continuous growth of the transformed cells, causing plant tumors. Rapid, cancerous growth of the transformed plant tissue obviously is advantageous to the bacterium: more food gets produced.

The Ti plasmid has been genetically modified ("disarmed") by deleting the genes involved in the production of bacterial food and of plant hormones, and inserting a gene that can be used as a selectable marker. Selectable marker genes generally are coding for proteins involved in breakdown of antibiotics, such as kanamycin. Any gene of interest can be inserted into the Ti plasmid as well. In principle, one can thus transform any plant tissue, and select transformants by screening for antibiotic resistance.

However, unfortunately, there are some complications:

- It has proven difficult to transform some monocots (grasses, etc.) by *Agrobacterium*, and
- Regeneration of plants from tissue culture or leaf discs is not always possible.

A number of genetically engineered plant varieties have been developed. Traits that have been introduced by transformation include herbicide resistance, increased virus tolerance, or decreased sensitivity to insect or pathogen attack. Traditionally, most of such genetically engineered plants were tobacco, petunia, or similar species with a relatively limited agricultural application. However, during the past decade it now has become possible to transform major staples such as corn and rice and to regenerate them to a fertile plant. Increasingly, the transformation procedures used do not depend on *Agrobacterium tumefaciens*. Instead, DNA can be delivered into the cells by small, μm-sized tungsten or gold bullets coated with the DNA.

The bullets are fired from a device that works similar to a shotgun. The modernized device uses a sudden change in pressure of He gas to propel the particles, but the principle of "shooting" the DNA into the cell remains the same. This DNA-delivery device is nicknamed "gene gun", and has been shown to work for DNA delivery into chloroplasts as well. Over the last several years, use of the "gene gun" has become a very common method to transform plants, and has been shown to be applicable to virtually all species investigated.

Transformation of rice by this method is now routine. This is a very important development as rice is the most important crop in the world in terms of the number of people critically dependent on it for a major part of their diet. Another method to get foreign genes into cereals is by electroporation: a jolt of electricity is used to puncture self-repairing holes in protoplasts (*i.e.*, the cell without the cell wall), and DNA can get in through these holes.

However, it is often very difficult to regenerate fertile plants from protoplasts of cereals. Nonetheless, significant advances in overcoming these practical difficulties have been made over the years. Now even transgenic trees have been created: the gene for a coat protein of the plum pox virus has been introduced into apricot. The plum pox virus leads to the feared Sharka disease, for which there is no cure. The resulting transgenic tree shows a markedly decreased sensitivity to this virus. The reason why continuous exposure of the tree to the viral coat protein leads to tolerance against viral infection is not yet understood, however.

Thus, now there are a number of different techniques to introduce foreign genes into plants. Essentially all major crop plants can be (and have been or are being) genetically engineered, the procedures are now routine and the frequency of success is very high. Even though genetically engineered crops are more costly than the usual ones, they have been rather readily accepted by US farmers provided that tangible benefits can be demonstrated. However, it is questionable whether the farmer in poorer countries can come up with the funds to "try out" and use the new crops.

Another issue in this respect is how genetically engineered crops are perceived by the consumer. Even though in the US there is little resistance to such crops as long as the products can be shown to be safe and advantageous, in other countries genetically modified foods are received poorly by the consumer. It is unlikely that there is a rationally sound basis for this rather hostile reaction of the consumer, as most of the crops are the result of human manipulation and may have been treated with harmful herbicides and pesticides.

Time and education will need to be invested to provide consumers and consumer advocates with a balanced opinion on the acceptability of the origin of their foods. One area of particular concern for some people is the lack of labeling of genetically engineered foods, and legislation may be introduced to address this issue.

On the other hand, as so many plants (soybean, corn, etc.) are genetically modified and the nature of the genetic modification is not necessarily easy to explain, it may be simpler to label those foods that are guaranteed free of "genetically modified organisms" or their products. However, keep in mind that essentially all agricultural products have been genetically modified by traditional breeding, so it may be difficult to define what is actually free of genetically modified organisms.

THE TOOLS OF GENETICS: RECOMBINANT DNA AND CLONING

To splice a human gene (in this case, the one for insulin) into a plasmid, scientists take the plasmid out of an *E. coli* bacterium, cut the plasmid with a restriction enzyme, and splice in insulin-making human DNA. The resulting hybrid plasmid can be inserted into another *E. coli* bacterium, where it multiplies along with the bacterium. There, it can produce large quantities of insulin.

In the early 1970s, scientists discovered that they could change an organism's genetic traits by putting genetic material from another organism into its cells. This discovery, which caused quite a stir, paved the way for many extraordinary accomplishments in medical research that have occurred over the past 35 years.

How do scientists move genes from one organism to another? The cutting and pasting gets done with chemical scissors: enzymes, to be specific. Take insulin, for example. Let's say a scientist wants to make large quantities of this protein to treat diabetes. She decides to transfer the human gene for insulin into a bacterium, *Escherichia coli*, or *E. coli*, which is commonly used for genetic research. That's because E. coli reproduces really fast, so after one bacterium gets the human insulin gene, it doesn't take much time to grow millions of bacteria that contain the gene.

The first step is to cut the insulin gene out of a copied, or "cloned," version of the human DNA using a special bacterial enzyme from bacteria called a restriction endonuclease. (The normal role of these enzymes in bacteria is to chew up the DNA of viruses and other invaders.) Each restriction enzyme recognizes and cuts at a different nucleotide sequence, so it's possible to be very precise about DNA cutting by selecting one of several hundred of these enzymes that cuts at the desired sequence. Most restriction endonucleases make slightly staggered incisions, resulting in "sticky ends," out of which one strand protrudes.

The next step in this example is to splice, or paste, the human insulin gene into a circle of bacterial DNA called a plasmid. Attaching the cut ends together is done with a different enzyme (obtained from a virus), called DNA ligase. The sticky ends join back together kind of like jigsaw puzzle pieces. The result: a cut-and-pasted mixture of human and bacterial DNA.

The last step is putting the new, recombinant DNA back into *E. coli* and letting the bacteria reproduce in a petri dish. Now, the scientist has a great tool: a version of *E. coli* that produces lots of human insulin that can be used for treating people with diabetes.

So, what is cloning? Strictly speaking, it's making many copies of a gene—in the example above, *E. coli* is doing the cloning. However, the term cloning is more generally used to refer to the entire process of isolating and manipulating a gene. Dolly the cloned sheep contained the identical genetic material of another sheep. Thus, researchers refer to Dolly as a clone.

Rules Governing RNA's Anatomy Revealed

"RNA is a very floppy molecule that often functions by binding to something else and then radically changing shape," said Al-Hashimi, who is the Robert L. Kuczkowski Professor of Chemistry and a professor of biophysics. These shape changes, in turn, trigger other processes or cascades of events, such as turning specific genes on or off.

Because of the RNA molecule's mercurial nature, "you can't really define it as having a single structure," Al-Hashimi said. "It has many possible orientations, and different orientations are stabilized under different conditions, such as the presence of particular drug molecules."

A major goal in structural biology and biophysics is to be able to predict not only the complex three-dimensional shapes that RNA assumes (which are dictated by the order of its nucleic acid building blocks), but also the various shapes RNA takes on after binding to other molecules such as proteins and small-molecule drugs. Further, researchers would like to be able to manipulate the 3-D structure and resulting activity of RNA by tweaking the drug molecules with which it interacts. But to do that, they need to understand the rules that govern the anatomy of RNA.

The quest has parallels to the study of human anatomy, Al-Hashimi said. "Your body has a specific shape that changes predictably when you are walking or when you are catching a ball; we want to be able to understand these anatomical rules in RNA."

Manipulating RNA is a much sought-after goal, given the recent explosion in vital cellular roles ascribed to RNA and the growing number of diseases that are linked to RNA malfunction. RNA performs many of its roles by serving as a switch that changes shape in response to cellular signals, prompting appropriate reactions in response. The versatile molecule also is essential to retroviruses such as HIV, which have no DNA and instead rely on RNA to both transport and execute genetic instructions for everything the virus needs to invade and hijack its host.

In earlier work, Al-Hashimi's team determined that rather than changing shape in response to encounters with drug molecules, RNA goes through a predictable course of shape changes on its own. Drug molecules simply "wait for" the right shape and attach to RNA when the RNA assumes the particular drug's preferred orientation, Al Hashimi said.

But what rules control the predictable path of shapes the RNA molecule assumes? And are those rules the same for all sorts of RNA molecules? In the current work, Al-Hashimi's team investigated those questions.

"RNA is very similar to the human body in its construction, in that it's made up of limbs that are connected at joints," Al-Hashimi said. The limbs are the familiar, ladder-like double helix structures, and the joints are flexible junctions. The prevailing view was that interactions among loopy structures at the tips of the limbs played a role in defining the molecule's overall 3-D shape, much as a handshake defines the orientation of two arms, but Al-Hashimi's group decided to look at things from a different perspective.

"We wondered if the junctions themselves might provide the definition," Al-Hashimi said. "If you look at your arm, you'll notice that you can't move it, relative to your shoulder, in just any way; it's confined to a certain pathway because of the joint's geometry. We wondered if the same thing might be true of RNA."

To investigate that possibility, the researchers turned to a database of RNA structures and found that all structures with two helices linked by a particular type of junction called a trinucleotide bulge fell along the same pathway.

The team then went on to explore structures of RNA molecules with other kinds of junctions. All were confined to similar pathways, but the precise pathway of a given RNA depended upon structural features of its junction. Just as anatomical features of our shoulders, elbows, hips and knees define the range of motion of our arms and legs, the anatomy of RNA's junctions dictates the motion of its helices.

Next, Al-Hashimi and coworkers wanted to understand how drug molecules cause RNA molecules to freeze in specific positions. In earlier work with an RNA molecule known as TAR, which is critical for replication of HIV and thus a key target for anti-HIV drugs, the researchers had found that certain drug molecules froze the RNA molecule in a nearly straight position, while others trapped the molecule in a bent conformation and still others captured positions between the two extremes. But because that project involved a wide variety of drug molecules, it was hard to figure out why certain ones preferred certain orientations.

To explore the issue more methodically, Al-Hashimi's group used a series of aminoglycosides (antibiotics that are known to target RNA) that systematically differed from one another in charge, size and other chemical properties. Size turned out to be the key: bigger aminoglycosides froze RNA in more bent positions; smaller ones favoured straighter RNA structures. Looking more closely, the researchers discovered that the aminoglycoside molecule nestles between two helices and acts like wedge, forcing the helices apart. Examination of other RNA structures bound to small molecules revealed that this rule is not specific to TAR but a more general feature of RNA-small molecule interactions.

"With these findings, it now should be possible to predict gross features of RNA 3-D shapes based only on their secondary structure, which is far easier to determine than is 3-D structure," Al-Hashimi said. "This will make it possible to gain insights into the 3-D shapes of RNA structures that are too large or complicated to be visualized by experimental techniques such as X-ray crystallography and NMR spectroscopy. The anatomical rules also provide a blueprint for rationally manipulating the structure and thus the activity of RNA, using small molecules in drug design efforts and also for engineering RNA sensors that change structure in user-prescribed ways."

DISCOVERY OF THE HYBRID HELIX AND THE FIRST DNA-RNA HYBRIDIZATION

Here we describe early research on RNA structure and the discovery of the DNA-RNA hybrid helix, a key component of information transfer. More than 50 years ago it was realized that the interaction between DNA and RNA was at the core of molecular biology. The problem was chemical in nature: could two different types of molecules interact and serve in the transmission of biological information?

In the 1950s it was widely assumed that "DNA makes RNA, RNA makes protein." This was not based on experimental evidence that DNA and RNA could combine but was more in the nature of an intuitive belief. However, by early 1960 I was finally able to carry out a direct experiment, the first DNA-RNA hybridization. In 1960, messenger RNA was still 1 year in the future, and there was not a great deal of understanding of the major components of information transfer. The DNA double helix proposed by Watson and Crick in 1953 clearly suggested that information was contained in the order of nucleotides, but during the 1950s our understanding of RNA was fragmentary.

The origins of the 1960 experiment go back to the mid-1950s. In 1954, while at Caltech, Jim Watson and I had been trying to find out if RNA by itself could form a double helix, but the fibre diffraction studies of RNA gave inconclusive results. The fuzzy diffraction patterns all looked alike, but unlike DNA, the RNA base ratios were all different in the samples we examined. It was clear that RNA was more complex than DNA, and its structure was unknown. In their 1953 paper Watson and Crick pointed out that it was probably impossible to form their double helix with ribose due to van der Waals interference of the 22-OH with the structure. Thus, it was likely to be different.

A NEW TYPE OF POLYMER CHEMISTRY

The research changed dramatically at the National Institutes of Health (NIH) in 1956 when David Davies and I began working with synthetic polyribonucleotides made using the polynucleotide phosphorylase enzyme discovered by Grunberg-Manago and Ochoa. When we mixed together polyriboadenylic acid (poly(rA)) and polyribouridylic acid (poly(rU)) and pulled fibres, they yielded a clear diffraction pattern of a double helix. The two molecules had combined to make an RNA double helix! This was a new type of chemical reaction, one that involved thousands of units binding specifically to each other in an extended array. Furthermore, no polymer reaction had ever been seen in which the monomers from two polymers bound together with great specificity. Two weeks after sending off the 1956

Journal of the American Chemical Society note, I wrote a letter to my postdoctoral mentor, Linus Pauling, describing these results. The letter reveals a sense of incredulity on my part that this reaction could happen and that it was "completely reproducible." The experimental demonstration that these molecules could form a double helix seems obvious today. However, it was a considerable surprise at that time and was often greeted with skepticism. Most biochemists felt that a double helix could only be made by an enzyme, such as the one discovered by Arthur Kornberg and associates, which appeared to replicate the DNA double helix. Polymer chemists felt that very long molecules involving thousands of nucleotides would probably become entangled and could not sort themselves out to form a regular double helix. Still other researchers, on theoretical grounds, felt that two highly negatively charged polymers were unlikely to combine to make a single structure.

However many scientists were receptive. In the early fall of 1956, a McCollum Pratt meeting was organized at Johns Hopkins University in Baltimore around the subject of the "Chemical Basis of Heredity." It was an excellent meeting with all of the major research workers in the field. In my talk I included a discussion of the specificity of the interaction and the fact that these long molecules seek each other out in solution and adopt this elongated helical form. Julian Huxley, a prominent scientist and writer, came up to me after my talk and warmly congratulated me for having discovered "molecular sex."

Gradually, the idea became fixed in the thinking of biologists and biochemists that these long nucleic acid molecules had considerable flexibility and could be made to form helical structures in solution. Although no one used the phrase at the time, this was the first hybridization reaction, and it represented a paradigm shift in the way chemists and biochemists thought about macromolecular nucleic acids.

Two Different Structures

Watson and Crick pointed out that RNA would not form the DNA structure. This was confirmed as the RNA x-ray diffraction pattern revealed significant differences from the DNA double helix pattern. The diameter of the RNA double helix was 6 Å larger than the DNA double helix. Analysis of the diffraction pattern revealed that the first layer line of the RNA-RNA duplex was stronger than the second—a reversal from that seen in the diffraction pattern of the DNA-DNA duplex. Another significant difference was that alterations in relative humidity changed the DNA double helix, producing both the A and B forms, whereas the RNA double helix seemed invariant to changes in relative humidity.

The reaction between poly(rA) and poly(rU) was associated with a drop in optical density in the ultraviolet, a property that could be used to identify the 1:1 stoichiometry and analyse the reaction quantitatively. Only much later was it realized that the RNA duplex structure was close to the dehydrated A conformation of DNA duplex. However, by 1960, it was clear that DNA and RNA duplexes were significantly different.

In 1957, together with Felsenfeld, we discovered that the poly(rA)·poly(rU) duplex could take on a third strand of poly(rU) to form a triple helix. We pointed out that this could be associated with forming two hydrogen bonds between the incoming uracil O4 and N3 and adenine N6 and N7, an interaction of bases that was confirmed 2 years later by Hoogsteen in a single crystal x-ray analysis. Between 1956 and 1960 a number of experiments were carried out with polyribonucleotides of different composition, and their interactions could generally be explained in terms of the ability of the bases to form at least two hydrogen bonds in the centre of the molecule.

Could DNA and RNA interact? Polymers of ribonucleotides were available because of the discovery of polynucleotide phosphorylase. This made it possible to produce polyribonucleotides with a variety of compositions, but there was no analogous method of producing DNA polymers. The question "How does DNA make RNA?" remained an open issue. Several biochemists were trying to isolate the enzyme known as RNA polymerase which is dependent upon a DNA template.

Experiments by Stevens, Hurwitz, and Weiss developed preparations that had some activity, but they were not purified enough to show what was actually happening. In a reflective article published in 1959 called "An Analysis of the Relation between DNA and RNA," I surveyed the various possibilities for DNA-RNA interactions. In particular, I asked whether RNA synthesis could be based on a double-stranded primer or a single-stranded primer of DNA. The discovery of several triple helical complexes of RNA molecules made it reasonable to consider a model in which RNA nucleotides were assembled by binding in a sequence-specific manner to double-stranded DNA. The analysis showed that such a model was unlikely because there were not enough stereospecific interactions to specify an RNA molecule. I concluded that it was likely to be based on forming an RNA strand on a single-stranded primer. These conclusions were fortified by the recent discoveries that denatured DNA could provide a primer for Kornberg's DNA polymerase enzyme, and furthermore, Sinsheimer had discovered that the virus ÕX174 had a single-stranded genome and it was a primer for the DNA polymerase. This reinforced the idea that single strands were adequate templates. In the same article, I speculated that RNA was likely to be the first polynucleotide molecule in the origin and early evolution of life and also pointed out the possibility of an enzyme that used a single-stranded RNA primer to make a DNA strand; it was called reverse transcriptase when it was discovered 10 years later.

Formation of a Hybrid Helix

Could a DNA-RNA hybrid helix form and have the stability needed so that it could be used for information transfer in view of the different physical properties and geometries of the RNA and the DNA duplex? I was finally able to address this problem with the chemical synthesis of oligodeoxythymidylic acid by Khorana and colleagues. Khorana kindly gave me a reaction mixture, which I fractionated and took the longer molecules to see if they would react with polyriboadenylic acid. The experiment worked. The discovery in 1960 that these could form a double helix represented the first experimental demonstration that a hybrid helix could be a method for the transfer of information from DNA to RNA. The evidence was based on measurements of the hypochromism, which occurs when helical molecules are formed, and the changes in sedimentation rate associated with complex formation. These results indicated that DNA could make RNA by using a single-stranded template to make a complementary RNA strand, and it was a model for RNA polymerase activity. However, the discovery of messenger RNA was still 1 year in the future. It is interesting that 1 year later in 1961, with a more highly purified preparation of RNA polymerase and using as a template the same oligodeoxythymidylate synthesized by Khorana, Hurwitz was able to synthesize polyriboadenylic acid using his enzyme preparation. This proved that a single-stranded template was adequate for RNA polymerase.

The publication in the summer of 1960 was the first demonstration of DNA-RNA hybridization although that term had not yet been invented. That particular hybridization is still widely used today in that immobilized oligo(dT) molecules are used to isolate eukaryotic messenger RNA through their poly(rA) tails. At the same time in 1960, Marmur, Doty, and colleagues found they could take denatured DNA molecules and hold them for a prolonged period at an intermediate temperature, called an annealing temperature, thereby allowing the DNA molecules to find each other and re-form a double helix. After presenting my work on DNA-RNA interactions at the Gordon Conference in 1960, Sol Spiegelman and Ben Hall came up to me and said they were inspired to try that reaction with a viral system. One year later they combined my work with the annealing work of Marmur and Doty and found that a newly synthesized RNA strand from T2 virus infection could be similarly annealed with DNA from the virus to make a hybrid helix. Thus, by 1961 both DNA-DNA and RNA-DNA hybridizations were available for a variety of biological studies.

The significance of the discovery of the DNA-RNA hybrid helix is relevant not only to understanding the activity of RNA polymerase, but it also explains how reverse transcriptase works, as well as telomerase, retrotransposons, and a variety of other interactions in which DNA and RNA strands combine. Today, fluorescence *in situ* hybridization is used in which DNA-RNA hybrids are formed to identify specific areas of the genome. Similarly, microarray gene expression profiling studies use DNA-RNA hybrid formation and are dependent on this 1960 discovery.

The full structural analysis of how these RNA and DNA chains accommodated each other was not revealed until much later. In 1982 the first DNA-RNA single-crystal hybrid molecule was solved in my laboratory. It showed that a DNA decamer containing four ribonucleotides formed a hybrid segment that stabilized the entire molecule in the A conformation. In 1992, we carried out a single-crystal x-ray analysis of an Okazaki fragment in which the nucleating ribonucleotides that initiate DNA synthesis in DNA replication were crystallized on DNA. Again, the presence of a few ribonucleotides was sufficient to convert the entire fragment into the A conformation.

The central role of the ribose ring puckers that differ in DNA and RNA is widely understood today. The van der Waals crowding of the ribose C22 oxygen determines the RNA pucker in the RNA double helical A conformation. This interaction provides a sufficient energy barrier to prevent a change in pucker in double helical RNA, in contrast to the facile manner with which the deoxyribose in DNA changes its ring pucker on lowering the water content to form A-DNA or to accommodate the presence of RNA in a hybrid double helix.

The hybrid DNA-RNA helix remains the bedrock of information transfer in biological systems. Indeed, the existence of a hybrid helix seems so obvious today that young research workers simply take it for granted. There is little realization of the extent to which, almost a half-century ago, scientists wrestled with problems of understanding how different polymers can react together to make a stable structure. The roots of our understanding of hybrid helix formation go back to 1960 and even further to the mid-1950s.

3

Gene Mapping of Plant

An important, though by no means an essential, step in genetic analysis is to produce genetic maps of the marker loci. Such maps are often referred to as 'framework maps' because they provide a framework within which important genes can be located, as well as providing a means of comparing chromosome organization in other closely or distantly related species. There are two stages to mapping. Firstly, to arrange the markers in a linear sequence separated by an appropriate map distance, *i.e.* to construct a linkage map. The second is to relate the linkage maps to particular recognizable chromosomes. The latter is often the most difficult and generally not essential for breeding or conservation work. We will, therefore, concentrate on the former. Chromosomes contain a single linear molecule of DNA and hence the markers on that chromosome occur at particular positions along that molecule. Typically, each chromosome contains 107 to 108 base pairs (bp), *i.e.* 104 to 105 kilo-base pairs (kbp) of DNA, while a typical structural gene, coding for a polypeptide chain, would be between 1 and 2 kbp long. Assuming only 10% of the genome is coding DNA, then a chromosome probably contains something of the order of 1000 to 10 000 genes. Fortunately, the fact that the chromosome is a linear molecule means that the genes need to be mapped in only one dimension.

Currently, the only useful method of gene mapping, at least as far as breeders and conservationists are concerned, relies on recombination between homologous chromosomes resulting from genetic exchange, which can be seen as chiasmata at diplotene through first metaphase of meiosis. A single chiasma on a particular chromosome results in half the gametes from such a meiosis being recombinant for that chromosome: half the gametes will contain a copy of that chromosome that is part maternal and part paternal in origin, while the remaining gametes will be entirely parental. Very few plant species have a sufficiently low number of clearly recognizable chromosomes that the number of chiasmata on a particular chromosome can be compared in different nuclei.

However, in many species the total number of chiasmata in each diplotene nucleus can be counted, at least in pollen meiosis, and the average number of chiasmata per nucleus calculated. The number and position of chiasmata on a particular pair of homologous chromosomes will vary from nucleus to nucleus; it is as though there are a large number of potential sites of exchange, but that in any given meiosis only a few sites are actually involved.

The longer the chromosome the more potential sites there might be. There will invariably be one chiasma and there may be two, three, four or more; a typical set of results is shown in Table A chromosome that has one chiasma on average is said to be 50 centimorgans (cM) long – a map length unit named after the American geneticist Thomas Hunt Morgan.

This relates to the fact mentioned above that one chiasma results in 50% recombinant chromosomes. By extension, a chromosome with an average of 2.5 chiasmata is said to be 125 cM (*i.e.* 2.5 350 cM) long. It appears that, as a rough rule of thumb, each chromosome has on average two chiasmata and so is 100 cM long. It follows, therefore, that providing the haploid chromosome number is known (*n*), the total map length will be approximately 2 3 *n* 3 50 cM. The average number of chiasmata per nucleus has been calculated in many species and this rule generally holds.

This rough rule is important because it gives the geneticist an idea of the total genetic map length that should be expected as more genes or markers are mapped. It provides a guide to the extent to which the currently mapped markers cover the full genetic map. Because the map is based in chiasma units, it is not the same as a physical map. The distribution of chiasmata is not uniform over a given chromosome in a species and will vary with the chromosome and the species.

Table. Chiasma Fraquency and Mapping.Chiasma Frequency in Chromosomes of Secale Cereale.

Number of chiasmata per chromosome	Frequency	Percentage unrecombined Chromosomes
1	0.267	50.0
2	0.716	25.0
≥3	0.017	12.5
Mean	–	22.0

Although a knowledge of the number of chromosomes and, ideally, the mean chiasma frequency per nucleus for the species, indicates the total map length, the actual map has to be constructed from examining the frequencies of progeny in crosses; *i.e.* maps are constructed from information based on the consequences of chiasmata, namely recombination of genetic markers.

Recombination of genetic markers is identified by the presence of gametes that contain recombined markers, and such gametes can be recognized from the phenotypes of progeny of a cross.

The frequency of such gametes is an estimate of the recombination frequency between the two markers concerned. The closer two markers are to each other on the chromosome, the less likely a chiasma will occur between them and the less likely they are to recombine. Providing the two markers are sufficiently close that either none or one chiasma occurs between them but never more, then the recombination frequency as a percentage is equal to the map distance in centimorgans as described above. The problem arises when the markers are sufficiently far apart for two or more chiasmata to occur between them in some meioses. It can be shown that not only does a single chiasma between a pair of markers result in 50% recombination between them, but so also do two, three or more chiasmata, on average. It is necessary to say 'on average', because there are a variety of possible consequences with two, three or more chiasmata in any given meiosis but, in practice, when one is looking at the progeny of a cross or self, each progeny will be the result of gametes from different meioses. This fact implies that although map distance increases linearly with the number of chiasmata, the same is not true with the frequency of recombination, which can never be more than 50% for a pair of markers even though they might be 100 cM apart at opposite ends of a chromosome.

In order to overcome this problem, mapping functions have been devised to correct for multiple chiasmata. The most common of these are the Haldane (1919) and Kosambi (1943) mapping functions. Haldane's assumes that the probability of none, one, two or more chiasmata in a given interval follows a Poisson distribution, *i.e.* that the chiasmata are independent, random events. Kosambi's method allows for a certain degree of nonindependence in chiasma occurrences. It has long been known by cytogeneticists that chiasma interference occurs over quite large regions of a chromosome.

Interference means that if a chiasma occurs at a particular point on the chromosome, the next one will never occur closer than some fixed distance, the interference distance, from it. Beyond this distance, there is a short distance in which the interference disappears and the next chiasma can then occur randomly outside this range. Observational data in various species of plants are confirming this and suggest that the interference distance is generally between 15 and 20 cM, *i.e.* 15 to 20% of a typical chromosome on either side of the chiasma. This has important and useful consequences for genetic and breeding work as will be shown later.

Recombination frequencies between pairs of markers can be scored in a wide variety of different crosses. The simplest to use are generations derived from an F1 because only two alleles are segregating, and their distribution in the chromosomes of the parents of the F1 can be determined. The generations that can be used are F2, backcrosses (Bc), recombinant inbred lines (RILs) or doubled haploid (DH) lines, with F2s being the most informative. General formulae for calculating recombination frequencies are given by Allard. In outbreeding species where F1s are not available, a given individual can be considered as an F1 and its selfed progeny as being an F2, even though the inbred parents do not exist. If it cannot be selfed but controlled crossing to another individual is possible, then again those genes segregating in the cross can be mapped although the situation is more complex.

There could be from two to four alleles segregating at each locus and, with respect to any one locus, the cross could represent an F2, Bc1 or Bc2. Moreover, the distribution of alleles in the two parents cannot be known and has to be inferred from the progeny. This complexity is compounded by the large number of marker loci that may be segregating in any given cross. Fortunately, a range of software packages are available to estimate these recombination frequencies, identify linkage groups, assign the markers to the most likely order and space them in map units (cM) on these linkage groups; examples are MAPMAKER and JOINMAP. Ideally the number of linkage groups should be equal to the chromosome number in the gametes but, unless the markers available provide a good coverage of the genome, it is likely that those markers on a given chromosome may appear as two or more separate linkage groups simply because the subsets are not sufficiently close for them to be recognized as being together on one chromosome.

The chromosomes of *Brassica oleracea*. Clearly these maps provide a very detailed coverage of the chromosome. Many of the markers are very close, *i.e.* less than 10 cM, and over this distance it matters little which of the two mapping functions is used. With the more widely spaced markers, *i.e.* recombination frequency >15%, Haldane's function will exaggerate their distance apart because it will allow for more double crossovers than actually occur.

As more markers are mapped, the total length should converge on the value predicted by the chiasma frequency, *i.e.* approximately 2 3 n 3 50 cM. It is important to remember that genetic map distances have standard errors which are dependent on the size and type of population used to construct the map, the mapping population. The map obtained is that which best fits, given the data and the assumptions underlying it, but there may well be other maps which also fit the data well, though are slightly less likely. Any particular map always develops a greater aura of respectability once it is published, frequently without stating its reliability, and subsequent yet different maps are treated with suspicion. A recombination frequency of true value p has a standard error of ?[p(1 2 p)/N], where N is the size of the gamete population. For example, if p is 0.1 (*i.e.* RF = 10%) and$N$ = 100, the standard error is 0.03 or 3%. In other words the 95% confidence interval of the estimate lies approximately between 4% and 16%.

There are other reasons, apart from statistical sampling, why maps could be wrong. The most obvious is the accuracy with which the data are scored and recorded. Autoradiographs and banding patterns on gels can easily be misread and research workers are loath to discard data even though its interpretation is ambiguous. All data should be checked independently by two or more people and any ambiguities removed.

Wrong scores will bias the data and often exaggerate map lengths as they suggest double recombination; wherever double recombination appears to have occurred in two short, adjacent intervals, the data should be checked. Changes in methylation rather than scoring errors could cause a restriction site to appear or disappear so giving the false impression of double crossing-over, or it could simply be an error.

Where maps appear to have major inconsistencies in different crosses, different chromosomal structural arrangements may be responsible, such as translocations or inversions. There is considerable evidence that chiasma frequencies, and hence recombination frequencies, may be quite different in male and female meioses even on the same plant although in other studies this may not be so. Thus although the order of markers on a linkage group may remain the same, the relative distances between them may be quite different if the map derives from male versus female meioses. For example, in *Brassica* it appears that the genetic map obtained from a backcross where the F1 is the female parent is 60% longer than when the F1 is the male parent. It is also clear that environmental factors during meiosis, particularly temperature, can affect the number and distribution of chiasmata, and this could be very critical if the crosses are set up at different times of the year.

Two or more different populations will almost certainly be segregating for different combinations of polymorphic markers, although some at least should be in common. These common ones can be used to overlay the two or more maps and a consensus map produced from an amalgamation of these. Again software is available to produce these consensus maps. Despite the various error-causing factors discussed above a considerable degree of consensus can be produced from such populations and, considerable similarity of chromosome sequence is conserved between even distantly related species, allowing the potential for cross-species genetic transfer.

It was stated above that the total map length should be dictated by the chiasma frequency, in so far as this can be accurately measured. When the first few genes were initially put onto genetic maps in the first 40 years of this century, they only covered a small part of the genome – except in wellstudied species such as *Drosophila* and maize. As more and more genes were placed on the map, so the total map lengths increased towards the asymptote dictated by the chiasma frequency. However, during the first decade following the use of molecular markers, the map lengths of some species appeared to exceed that expected, and it was even argued by some that the chiasma theory of recombination might be wrong.

No one would wish to argue that chiasma frequency data are free from error but there would have had to have been quite excessive underscoring to result in the disparity which was occasionally found. It would appear, however, that the excessive lengths were due in part to errors in scoring and to the use of small mapping populations, because subsequent map sizes have shown a progressive decrease towards the predicted asymptote. Scoring errors bias estimates of RF generally upwards, while RFs have to be large to be detectable in small populations. Our own experience with mapping in cereals and brassicas has also shown a progressive decrease in estimated map length as more markers can be found to fill some of the large gaps in the map and as errors are removed from the mapping data.

Gene Cloning

Providing a target gene can be located to within a pair of marker loci, it should be possible to use standard cloning techniques to walk along the chromosome between the markers and to find the target gene. The feasibility of this depends on the distance between the markers and the ease with which the target gene can be recognized from among the clones.

Depending on the species, a genetic length of 1 cM can, on average, consist of as few as 280 kbp of DNA in rice, or 2220 kbp in maize. It is not always easy to locate a target gene with sufficient accuracy to be able to say that it is between two markers as close as 1 cM, and hence the actual lengths of DNA between them can often be much greater. With QTLs one can seldom achieve anything approaching this accuracy by existing methods of conventional mapping. Furthermore, these are average distances per centimorgan; some chromosomal regions could be much longer or, indeed, less.

Increases in mapping reliability can be obtained by using what are called near isogenic lines (NILs), which are lines obtained by selfing or backcrossing and are known to differ from some standard genotype by just a short defined section of chromosome. Providing this section is delineated by accurately mapped flanking markers, and the NILs which do and do not contain this region can clearly be shown to differ for the target gene, then the target gene has to be in that region. The genetic and physical proximity of the markers can be established by conventional mapping and by gene cloning.

Map-based gene cloning involves starting from one of the markers and walking to the next by means of a series of partially overlapping cosmid clones. Cosmid clones are necessary because they can be up to 50 kb in length and hence require fewer cloning steps in order to cover the DNA between the markers (280 to 2200 kb per cM in the examples cited above). Identifying the target gene is more difficult. Clearly, if the gene is between the two flanking markers it has to be on one of the cloned regions and could possibly be on two overlapping clones.

It could be identified by transforming plants with each clone separately and identifying which transformant expresses the effect. If the gene concerned exhibits classical dominance it would be necessary to transform the homozygous recessive with the dominant allele or the transformant would not be recognized. Alternatively, the gene has to be introduced into a species that does not normally express the gene. Once the effective cloned fragment is identified, the actual location of the gene within the fragment could be identified by successively transforming sub-fractions of the clone or, following sequencing, by looking for potential open reading frames. In those cases where the target gene produces a known product, cDNA clones derived from tissue likely to be rich in the appropriate mRNA of the target gene can be used to identify likely sequences in the previous overlapping clones.

Once the gene has been identified by one of these methods, its structure and activity can be studied in detail. Disease resistance genes are obvious candidates for such studies as their action is very specific and their location on the genome of many species is well documented. However, unless suitable candidate loci.

Methods for Cloning Plant Traits

Many of the first plant traits to be cloned were those in which some degree of biochemical understanding preceded any work with DNA. The understanding may have come from plant systems or from outside the plant kingdom. Most of the cloned genes in these cases were enzymes which were part of a well-characterized biochemical pathway or had a protein structure which was otherwise well known, such as seed storage proteins. Cloning genes based on a knowledge of the protein sequence can be achieved by working backwards from the gene product (amino acid sequence) to the coding sequence or the DNA template. Such an approach has been used for cloning many seed storage proteins, such as the 2S sulphur- rich proteins from Brazil nut.

Genes of plant origin have been used in transgenic plants to improve resistance to insect pests. One of the first to be characterized was the cowpea trypsin inhibitor, a protease inhibitor which acts as an antifeedant to a range of insect pests. The protein is naturally produced in seeds of some genotypes of cowpea (*Vigna unguiculata*), which was shown to confer lepidopteran insect resistance when expressed constitutively in tobacco. Another such report from the same group at Durham

University described the use of a snowdrop lectin gene to confer aphid resistance in transgenic tobacco, the first report of genetically engineered resistance to a sucking insect in plants. The lectin from snowdrop (*Galanthus nivalis*), known as GNA, has been demonstrated to be effective against two major rice pests, the brown planthopper and the green leafhopper, as well as the peach potato aphid. When expressed at 0.1% of total leaf protein in tobacco, GNA is antimetabolic to homopteran pests. These levels of protein were accumulated in leaf tissues when placed under the control of the CaMV 35S promoter. However, Hilder *et al.* (1995) postulate that better control would be achieved with the use of a phloem-specific promoter, such as the sucrose synthase 1 promoter from rice.

Another significant group of genes which have been cloned based on prior knowledge of the end product are the seed storage genes, including both seed storage proteins and lipids. In both cases there was considerable biochemical knowledge of the pathways involved in biosynthesis and storage. For seed storage proteins, the genes can be cloned based on the knowledge of the amino acid sequence of the mature protein and targeting signal or transit peptide.

Molecular genetic manipulation of vegetable oils is a major area of research in the developed world. While approximately 90% of vegetable oils are used for human consumption, there are emerging markets for industrial applications including lubricants, paints, cosmetics, plasticizers and soaps. With increasing emphasis on quality and dietary considerations, there is a desire in developed countries to substantially alter the types and balances of fatty acids found in the major oil crops. Six crop species (soybean, oil palm, rapeseed/canola, sunflower, cottonseed and peanut) account for 84% of the world's vegetable oil production, with the major fatty acids produced being palmitic, stearic, linoleic, linolenic and oleic acids.

However, almost 200 different fatty acids are produced by plants, mostly occurring in nondomesticated plant species. Hence, genetic manipulation of the major crop species, either via conventional breeding or genetic engineering, has the potential to modify the fatty acid types produced and stored by seeds such as soybean and canola.

The biochemistry of fatty acid biosynthesis is well established for many of the major pathways, with enzymes identified which control the chain length of fatty acids, and the degree of saturation, or number of double bonds in the fatty acid chain. The oils stored in seeds have been altered by downregulation of particular enzymes using antisense RNA techniques.

The oleic acid content of rapeseed oil has been increased to 83% (from 62% in conventionally bred cultivars) by downregulating the enzyme D12-oleate desaturase using antisense RNA. A similar result has been achieved using co-suppression, which has produced rapeseed oil with 87% oleic acid content. Genetic engineering approaches have also been used to introduce totally novel fatty acids to major oilseed crops. Petroselinic acid, an isomer of oleic acid, is found in abundance in the spice plant, coriander (*Coriandrum sativum*). This oil has uses in cosmetics and pharmaceuticals, and its oxidation leads to the formation of lauric acid, used in detergents, and adipic acid, which is used in nylon production.

Hence this is a potentially valuable oil which may enable soybean and canola crops – *i.e.* renewable resources – to replace some of the nonrenewable carbon fuels which are currently used for many industrial polymer and lubrication applications. A cDNA clone from coriander encoding an acyl-ACP desaturase has been genetically engineered into tobacco, resulting in the accumulation of petroselinic acid. Oil crops such as rapeseed and soybean could be genetically modified to produce this fatty acid on a commercial scale.

Hence, here is an example whereby a minor crop plant has become a donor of genetic diversity which will enable the commercial production of a very useful array of products for food and industrial purposes.

Cloning Genes Known Only by Phenotype

However, a more challenging task has been to move beyond proteins and pathways about which sufficient biochemistry is known into cloning genes known by phenotype only. By far the majority of genes fit into this category, which includes many of the most important plant traits such as disease resistance, adaptation to abiotic stresses, phenology and flowering behaviour and fertility. Only in the past few years have genes for traits known only by phenotype been cloned. One of the most interesting areas of plant biology is the understanding of plant–pathogen interactions at the molecular level.

The major methods used for cloning genes known only by phenotype are:

- Transposon tagging.
- Map-based or positional cloning.
- T-DNA tagging.

There have been a number of plant genes cloned using all these methods or modifications of them. There are also examples where a combination of methods has been used to expedite the rate of identification of the unknown sequence.

Transposon Tagging

The proposal that specific pieces of DNA are able to autonomously excise from a particular locus and insert, theoretically at random, into another linked or nonlinked locus was first proposed by Barbara McClintock. Transposable elements are now accepted as ubiquitous, and are particularly well characterized in organisms such as *Escherichia coli*, *Drosophila* and maize. In two plant species, maize and the snapdragon (*Antirrhinum majus*), transposable elements are sufficiently well characterized to be used for mutagenesis and gene tagging. The best understood system is the maize *Ac/Ds* (*Activator/Dissociation*) family of elements. The behaviour of the *Ac* element is such that it can excise autonomously from its locus, under the control of its own transposase, and insert into another chromosomal position.

Where the *Ac* element is in or near a gene, that gene is inactivated (*i.e.* not expressed), and will appear phenotypically as a recessive mutation. When *Ac* transposes (excises and reinserts at another locus), the gene is commonly returned to normal function, and hence is termed a revertant.

Should this *Ac* element then insert into another gene (transcribed or regulatory sequence), then a new recessive mutation will manifest. Transposition can occur either germinally, which results in an altered phenotype for the whole plant, or somatically, which results in a mosaic mutation, and can often be detected as a sectoring of plant or seed colour. The *Ds* element is actually a group of elements derived from defective *Ac* elements, with a common feature being their lack of effective *Ac* transposase.

Hence these elements are nonautonomous, and will only transpose in the presence of an *Ac* element capable of producing transposase. Equipped with sequence and map information of an *Ac* element in maize, geneticists have been able to clone important maize genes using the transposon-tagging approach.

Table. Plant Transposable Elements which have been used for Transposon-Taggine

Element	Species	Type of Element	Size
Ac	Maize	Autonomous	4.6
Ds	Maize	Nonautonomous	0.5–30
Tam1	Snapdragon	Autonomous	15
Tam2	Snapdragon	Nonautonomous	5
En/Spm	Maize	Autonomous	8.3

When a gene is inactivated by an *Ac* element, proof of which can be demonstrated by a low frequency of trait reversion, the *Ac* element can be used as a probe for 'fishing-out' the tagged gene. A tagged gene can be cloned by performing Southern blot analysis with the *Ac* element used as probe, and any polymorphism represents an excision/insertion event. The technique is described more fully in the review by Walbot.

The *Ac* system has proven extremely useful for cloning maize genes for traits such as plant pigment genes (*bronze*; the anthocyanin pathway), starch biosynthesis (*opaque2*), and abscisic acid (AbA) insensitivity. Other maize transposable elements such as *Mutator* (*Mu*) and the *En/Spm* system have been used to clone maize genes, as has the *Tam* system in snapdragon. As a result, the search began for transposable element systems in other species so that a similar approach could be taken.

However, it was soon demonstrated that the well-characterized *Ac/Ds* system could be genetically engineered into other plant species including tobacco, *Arabidopsis*, tomato and petunia. It was subsequently shown that the *Ac* element acted in a similar manner in these heterologous species, undergoing germinal and somatic excision and insertion, and maintaining the propensity to reinsert at a linked site (usually within 4 cM).

As a result, the frequency of mutation in the surrounding area approaches 1024. It was not until 1993 that a plant gene was cloned using the heterologous system, that of the *Ph6* gene in petunia, which is involved in floral pigmentation. Using the transposon-tagging approach, a number of plant disease resistance genes (R genes) have now been cloned. The first of these was the tomato *Cf-9* gene, which confers resistance to race 9 of the fungal pathogen *Cladosporium fulvum*, causal agent of leaf mould. The gene was tagged using a modified *Ac/Ds* system.

The system relied on the use of a stabilized *Ac* element (s*Ac*), which was effectively a disabled *Ac* element that nevertheless produced *Ac* transposase and hence enables *Ds* transposition. The two elements are maintained in separate lines, and hence are both stable until controlled crossing is performed.

This gives a significant advantage in that *Ds* insertion mutants can be maintained indefinitely in the homozygous or hemizygous form, and revertants can be isolated by crossing lines with s*Ac* tomatoes. The transposon-tagging approach used to isolate the *Cf-9* gene, which was known by phenotype only, did have considerable advantages in that much was known of the pathogen.

The hypersensitive response elicited by the virulent strain fitted into the gene-for-gene hypersensitive response proposed by Flor, with a dominant avirulence gene (*Avr9*) interacting with the *Cf-9* gene to form an incompatible or hypersensitive response. *Avr9* had previously been cloned and fully characterized, and was known to specify a 28 amino acid peptide which alone could elicit a necrotic response in resistant tomatoes.

Crosses were made to develop a transgenic tomato line homozygous for *Cf-9*, but heterozygous for *Ds*. The most seedlings died from systemic hypersensitive response, with *Avr9* and *Cf-9* present in all cells. The only survivors would be individuals where *Ds* had inserted into the *Cf-9* gene. Where this happened in plants carrying s*Ac*, the seedlings were variegated for the hypersensitive response (somatic excision), and in the absence of s*Ac*, the survivors were stable (germinal excision).

This work was a major undertaking, involving the germination of approximately 160 000 progeny, yielding 118 survivors, representing 63 independent mutations. After considerable effort to characterize these, the *Cf-9* gene was cloned and characterized, with an open reading frame encoding an 863 amino acid protein.

Map-based Cloning

The previously described transposon-tagging scheme utilized to clone *Cf- 9* had an element of the map-based approach by utilizing the phenomenon of *Ac*/*Ds* transposition to linked sites, with a *Ds* element within 3 cM of the locus of interest, which greatly enhanced the likelihood of insertional mutagenesis. However, map-based cloning requires a much tighter genetic linkage (in centimorgans or recombination fraction), which in turn needs to be converted into a physical distance (kbp of DNA sequence).

The map-based, or positional cloning concept relies on building a well-saturated genetic map of DNA sequences of the organism in question. The genetic map is built up in a polymorphic mapping population, commonly an F2, backcross (BC) or recombinant inbred, with markers such as RFLPs (restriction fragment length polymorphisms). The first RFLP map to be made was of human. Based on such maps, important human genes responsible for disorders such as cystic fibrosis and muscular dystrophy have been cloned in this manner.

Genetic maps of plant species have been made with DNA markers in virtually all important plant species, particularly the major cereals, grain legumes, oilseeds and important vegetable crops including tomato and potato. The RFLP marker system is based on Southern blot analysis, while many of the newer technologies use the polymerase chain reaction (PCR).

These include random amplified polymorphic DNA (RAPDs), amplified fragment length polymorphisms (AFLPs), and a number of different marker systems based on simple sequence repeats (SSRs), which may use PCR, such as inter simple sequence repeat (ISSR)-PCR, or Southern analysis. The applications of DNA markers to plant genetic analysis and breeding are many, and the subject has been reviewed by various authors. The population for saturation mapping should be segregating in a Mendelian manner for the trait of interest. This requires a lot of tedious but demanding work, and it may be that particular populations such as near isogenic lines, or DNA pooling strategies, such as bulked segregant analysis, will allow the mapping effort to be more targeted to the region of interest, hence improving the rate of map saturation.

It is desirable to obtain extremely close genetic linkage to the locus of interest (<1 cM), with flanking markers if possible. It can be of considerable advantage to identify a marker which co-segregates (say within 0.1 cM) with the trait. The next step is to convert the genetic map into a physical map of the chromosome region, translating distances from centimorgans to nucleotides.

The Gene of Interest (in this illustration,R) is Flanked by Markers such as RFLPs or SSRs. A Series of Overlapping YAC Contigs is used to cover th Chromosomal Region Containing the Gene.The YAC Which Contains the Complete Gene is Marked in Bold.This is then Subcloned or Matched to a cDNA Library to Idenfify the Gene(s) Contained the Region, and Ultimately Identify the R Gene.

Physical mapping then involves the use of YAC or BAC clones which span the region of interest. These YAC/BAC contigs will span the region between two flanking DNA marker sequences, a process known as chromosome walking. In this example, two YAC/BAC clones contain the gene of interest. Methodology is being developed to enable BAC clones to be directly transformed into plants, to find which one contains the gene of interest (in the example, the *R* gene). Alternatively, the YAC/BAC must be subcloned (for example as a cDNA library), and these subclones can be put into the susceptible plant and tested for the R phenotype.

The first R gene to be cloned in plants via map-based cloning was the *Pto* gene, which confers resistance to *Pseudomonas syringae* pv. *lycopersici*, the causal agent of bacterial speck in tomato.

The starting points for the work were:

- A high-density RFLP map of tomato.
- A tomato YAC library.

The *Pto* locus had been previously mapped to tomato chromosome 5 and one RFLP marker co-segregated with *Pto*. Hence this was a useful starting point to screen the YAC library. A 400 kb YAC was identified, and via inverse PCR, some end-specific probes were generated. The left arm of the YAC was 1.8 cM from *Pto*, with the right arm perfectly co-segregating. This YAC was then used to screen a 920 000 plaque leaf cDNA library, from which 30 were selected. One of these was shown to co-segregate with *Pto*. This cDNA was then transformed into a susceptible tomato cultivar under the control of the CaMV 35S promoter, where it conferred resistance. Hence the gene was identified, and could then be sequenced and characterized.

The success of this approach was partly due to the small genome size of tomato (950 Mbp), and a relatively low level of repeat sequence. Such an approach is much more difficult in species with larger genomes such as maize (2300–2700 Mbp) and wheat (16 000 Mbp). One such example of this problem is the 79 000 clone maize YAC library produced by Zeneca Seeds. As reported by K. Edwards, only 15% of the YACs have unique ends, with the remainder terminating in repeat sequences, which makes chromosome walking very difficult. The synteny of genomes is a major advantage in this instance. It is now well established that there is a high degree of synteny of genome organization, as seen in the co-linearity of maps of the most important grass genomes. This has resulted in the great interest in using this co-linearity to facilitate cloning of genes in large genomes such as maize and wheat, by starting with the smaller genomes, particularly rice, and to a lesser extent, sorghum.

One particular study of maize YACs demonstrated that a single YAC with the *Adh1* sequence contained 36 different repetitive sequences, whereas these repetitive sequences were largely absent in corresponding areas of the rice and sorghum genomes, a pattern seen also for the *a1–sh2* region in all three genomes. In a similar manner, many genes of interest for *Brassica* improvement are being identified in the smaller *Arabidopsis thaliana* genome.

Paterson *et al.* have recently demonstrated the potential of this approach, by identifying quantitative trait loci (QTLs) for seed mass and phenology which were co-linear on molecular marker maps of maize, sorghum and rice. They proposed that, as the maize genome is four and six times larger than the sorghum and rice genomes, respectively, the genes of interest in maize may be cloned in sorghum/rice more expediently. It must be remembered that QTLs are subject to genotype 3 environment interaction, which will be a major complicating factor in attempts to clone and characterize a QTL.

T-DNA Tagging

When T-DNA is inserted into the plant genome, it does so via illegitimate recombination. In the same way as a transposon, the insertion of T-DNA into a gene will inactivate it, which phenotypically will appear as a mutation. The comparative advantage of such a method is that stable mutations will be conferred. This can be a disadvantage, however, as reversions, such as those made possible by the *Ac*/*Ds* system are not possible. It is also the case that each independent transformation event will result in just one mutation event (or more if multiple independent insertions are made), whereas the heterologous transposon approach allows multiple mutation events to take place from a single initial transformant. Hence, mutated genes with T-DNA insertions can be cloned, in much the same manner as those with transposon insertions, by using part of the T-DNA as a probe in Southern blots.

The procedure for selecting out the tagged gene is quite straightforward, particularly when the plant contains a single T-DNA copy. However, this means that a large number of independent transformants must be generated. This can be achieved with species such as *Arabidopsis*, where Feldmann demonstrated that a seed-based transformation system using *Agrobacterium* can be quite efficient. Feldmann demonstrated that by inoculating approximately 1000 seeds, approximately 300 000 selfed T2 seed can be collected. They estimated that approximately 8000 transformants could be recovered from an experiment of this scale and, using this approach, cloned a gene involved in trichome development and a dwarf gene from *Arabidopsis*.

A regulatory gene in the ethylene biosynthesis pathway of *Arabidopsis* has also been cloned using the T-DNA tagging strategy. With modification, T-DNA tagging methodology can be used to clone plant promoter sequences. Koncz *et al.* designed a vector which contained a promoterless antibiotic selection gene (aminoglycoside phosphotransferase II) and transformed this into *Nicotiana* and *Arabidopsis*. In this way, promoters could be 'trapped' as plants that were regenerated on selective antibiotic were antibiotic resistant because the T-DNA was inserted downstream of a promoter such that the gene was expressed.

The most serious limitation to the T-DNA tagging approach is that large numbers of independent transformation events must be generated, which is not currently possible with most species. This is really only feasible with a system such as the seed transformation technique developed for *Arabidopsis*, or possibly the 'whole-leaf ' transformation system developed for tobacco.

TRANSPOSITION

Transposition is also a major cause of plant genome expansion. To begin with, transposition generates DNA. Retrotransposition results from transcription of genomic retrotransposons, followed by insertion of reverse transcripts into the genome at new sites . Plant transposons generate additional copies of themselves by virtue of excising from only one of two newly replicated sister chromatids and reinserting into as yet unreplicated sites . Absent countering forces, genome expansion is an inevitable consequence of the properties of transposable elements. The accumulation of retro- transposon blocks between genes is a major factor in the size difference between the maize genome and those of its smaller relatives (SanMiguel *et al.*, 1996, 1998).

Retrotransposon blocks occupy 74% of the recently sequenced 240-kb maize *Adh* region (Tikhonov *et al.,* 1999). These blocks contain 23 members of 11 different retrotransposon families, primarily as complete retrotransposons, but also occasionally as solo LTRs (Tikho- nov *et al.,* 1999). Within these blocks, retrotransposons are commonly nested by insertion of retrotransposons into each other (SanMiguel *et al.,* 1996, 1998).

What is perhaps most surprising about the maize retrotransposon blocks that have been characterized is that they grow quite slowly. The transposition mechanism assures that retrotransposon ends are almost always identical when an element inserts, hence the divergence between the LTRs of a single element reflects the age of the insertion. Bennetzen and his colleagues found that the sequence difference between the LTRs of a given element is almost invariably less than the sequence difference between the LTRs of the element into which it is inserted. Using these differences to order and date the insertions, they inferred that all of the insertions have occurred within roughly the last 5 million years, well after the divergence of maize and sorghum (SanMiguel *et al.,* 1998). Importantly, no retrotransposons have been found in the corresponding *Adh1* flanking sequence in sorghum (SanMiguel *et al.,* 1998; Tikhonov *et al.,* 1999). This raises the possibility that retrotrans- poson activity may differ between closely related lineages.

AMPLIFICATION AND REARRANGEMENT

New copies of transposons and retrotransposons provide new sites of homology for unequal crossing over. Evidence that transposable elements are central to the evolutionary restructuring of genomes has accumulated in every organism for which sufficient sequence data exist. Exceptionally detailed examples of the role of transposition, retrotransposition, amplification, and transposon-mediated rearrangements in the evolution of a contemporary chromosome are provided by recent studies on the human Y chromosome (Saxena *et al.,* 1996; Schwartz *et al.,* 1998; Lahn and Page, 1999a, b). Although the level of resolution is not yet sufficient in many cases to determine the molecular history of each duplication, it is evident that many, if not a majority of plant genes belong to gene families ranging in size from a few members to hundreds (Michelmore and Meyers, 1998; Riechmann and Meyerowitz, 1998; Martienssen and Irish, 1999; Rabinowicz *et al.,* 1999). *R* genes, for example, comprise a superfamily of similar *myc*-homologous, helix–loop–helix transcriptional activators of genes in anthocyanin biosynthesis (Ludwig *et al.,* 1989; Perrot and Cone, 1989; Consonni *et al.,* 1993). Detailed analysis of the *R-r* complex, a well-studied member of the *R* superfamily, reveals a history of transposon-catalyzed rearrangement and duplication . There also may be other genetic mechanisms that drive genome expansion. A recent analysis of the behavior of maize chromosomal knobs reveals that the pattern of segregation under the influence of a "meiotic drive" locus of as yet unknown function results in the preferential transmission of chromosomes with larger knobs over chromosomes with smaller knobs (Buckler *et al.*, 1999). Maize knobs are blocks of similar short tandemly repeated sequences, ranging from as few as 100 copies to as many as 25,000 per site (Ananiev *et al.*, 1998a). Their structure and dispersed occurrence further suggest that they are transposable (Ananiev *et al.*, 1998a, b; Buckler *et al.*, 1999). The combination of transposability and preferential transmission of chromosomes with expanded knobs thus provides an additional mechanism for genome expansion.

GENOME CONTRACTION

Are there genetic mechanisms that contract genomes? Careful analysis of the relative deletion frequency and length in drosophilid non-LTR retrotransposons supports the inference that there are more deletions per point mutation in Drosophila than in mammals and that the average deletion size is almost eight times larger (Petrov and Hartl, 1998). Thus mechanisms that contract genomes by preferential deletion may exist, as well. Bennetzen and Kellogg have argued that despite ample evidence for the operation of mechanisms that expand genomes in plants, there is little evidence that plant genomes contract (Bennetzen and Kellogg, 1997). The maize intergenic regions that have been analyzed, for example, comprise predominantly intact retrotransposons, rather than solo LTRs, which can arise by unequal crossovers between the repeats at retrotransposon ends and are common in other genomes (Bennetzen and Kellogg, 1997).

However, it also is known that both the Ac and Spm transposons of maize frequently give rise to internally deleted elements, and Ac ends are very much more abundant in the maize genome than are full-length elements, suggesting deletional decay of transposon sequences (Fedoroff *et al.*, 1983, 1984; Schwarz-Sommer *et al.*, 1985b; Masson *et al.*, 1987). So it would not be surprising to find mechanisms that preferentially eliminate sequences. And indeed, preferential loss of nonredundant sequences early after polyploidization has been detected in wheat (Feldman *et al.*, 1997).

CONTROLLING TRANSCRIPTION, RECOMBINATION, AND TRANSPOSITION

Despite our growing awareness of the abundance of plant transposable elements and the role they have played in shaping contemporary chromosome organization, the fact is they eluded discovery for the first half century of intensive genetic analysis. Thus what is perhaps the most striking observation about transposable elements is not their instability, but precisely the opposite: their stability. Not only are insertion mutations in genes infrequent, but retrotransposition events are so widely separated that the time interval between insertions in a particular region of the genome can be counted in hundreds of thousands to millions of years (SanMiguel *et al.,* 1998). Chromosomes containing many hundreds of thousands of transposable elements are as stable as chromosomes containing few. By what means are such sequences prevented from transposing, recombining, deleting, and rearranging?

The transposon problem can be viewed as one aspect of a larger problem in genome evolution: why does duplicated DNA persist? Duplications are a by-product of the properties of the DNA replication and recombination machinery. Short stretches of homology suffice to give rise to duplications by slippage during replication, homology-dependent unequal crossing-over, and double-strand breakage/repair (Gorbunova and Levy, 1997; Liang *et al.,* 1998). But duplications are problematical. Once a duplication exists, the mechanisms that generated it also permit unequal crossing over between identical repeats (Anderson and Roth, 1977; Perelson and Bell, 1977; Koch, 1979). Prokaryotes readily duplicate genetic material, but do not retain duplications (Perelson and Bell, 1977; Romero and Palacios, 1997).

Thus the ability of genomes to expand by duplication is predicated on their ability to sequester homologous sequences from the cell's recombination machinery and retain them, which may necessitate the invention of mechanisms to recognize and differentially mark duplications. Some lower eukaryotes, including *Neurospora crassa* (Selker and Garrett, 1988; Selker, 1997) and *Ascobolus immersus* (Rossignol and Faugeron, 1994, 1995), have the capacity to recognize and mark duplicated sequences by methylating them. Sequence methylation silences transcription, enhances the mutability of the duplicated sequence, and inhibits recombination (Selker, 1997; Maloisel and Rossignol, 1998).

Some years ago, Adrian Bird pointed out that there are two evolutionary discontinuities in the average number of genes per genome (Bird, 1995). The first is an increase between prokaryotes and eukaryotes and the second is between invertebrates and vertebrates. He suggests that with a given cellular organization there may be an upper limit on the tolerable gene numbers imposed by the imprecision of the biochemical mechanisms controlling gene expression. He suggested that the transcriptional "noise reduction" mechanisms that arose at the prokaryote/eukaryote boundary were the nuclear envelope, chromatin, and separation of the transcriptional and translational machinery, as well as RNA processing, capping, and polyadenylation to discriminate authentic from spurious transcripts. He proposed that genome-wide DNA methylation is the novel "noise reduction" mechanism that has permitted the additional quantal leap in gene numbers characteristic of vertebrates.

HOMOLOGY-DEPENDENT GENE SILENCING

The results of both classical and contemporary studies on the silencing of redundant gene copies in plants suggests that both methylation and other epigenetic mechanisms reflect a much more fundamental ability to recognize and regulate gene dosage . Mc- Clintock understood that transposable elements exist in a genetically intact, but cryptic form in the genome and she carried out genetic analyses of *Spm* transposons undergoing epigenetic changes in their ability to transpose . We later found that the genetically inactive *Spm* transposons are methylated in critical regulatory sequences (Banks and Fedoroff, 1989). It also has been reported that the large intergenic retrotransposon blocks in maize are extensively methylated (Bennetzen *et al.*, 1994).

The discovery that the introduction of a transgene can lead to the transcriptional silencing and methylation of both the introduced gene and its endogenous homolog brought gene silencing mechanisms under intense study (Park *et al.*, 1996; Kooter *et al.*, 1999). Genes can be silenced both transcriptionally and posttranscriptionally consequent on the introduction of additional copies. Posttranscriptional silencing appears to be caused by RNA destabilization, whereas transcriptional gene silencing involves DNA methylation (Vaucheret *et al.*, 1998; Kooter *et al.*, 1999). There is also some evidence that posttranscriptional silencing triggers DNA methylation (Wassenegger *et al.*, 1994).

The results of recent studies on the classical epigenetic phenomenon of *R* locus paramutation in maize have revealed that local endoredu- plication of a chromosomal segment both triggers silencing and can render the endoreduplicated locus capable of silencing an active allele of the gene on a homolog (Kermicle *et al.*, 1995). Similar observations have been made with transgenes, as well as endogenous gene duplications at different chromosomal locations in tobacco and *Arabidopsis* (Matzke *et al.*, 1994; Luff *et al.*, 1999).

A connection between gene silencing and chromatin structure has come from the analysis of mutants altered in methylation and in transcriptional gene silencing (Jeddeloh *et al.*, 1998; Kooter *et al.*, 1999). Both approaches have identified alleles of the *ddm1* locus, which encodes a protein with homology to known chromatin remodeling proteins. This suggests that the repressive mechanisms of DNA methylation and chromatin structure are linked in plants, as they are in animal cells (Ng *et al.*, 1999; Wade *et al.*, 1999). Evidence also is accumulating that doublestranded RNA mediates gene silencing, both in plants and in a variety of other organisms (Waterhouse *et al.*, 1998; Fire, 1999). Analyses of mutants altered in posttranscriptional gene silencing in *Neurospora* have identified an RNA-dependent RNA polymerase, as well as a RecQ helicase-like protein, homologs of which are known to be involved in DNA repair and recombination (Cogoni and Macino, 1999a, b).

THE ORIGIN OF TRANSPOSONS AND METHYLATION

Although it is popular to assert that transposons are genomic "parasites" and that DNA methylation evolved to control them, I suggest that the evidence supports neither notion (Yoder *et al.*, 1997). The idea that transposons as parasitic, selfish DNA comes from a couple of essays written two decades ago, one by Doolittle and Sapienza (1980) and one by Orgel and Crick (1980). These essays sought rightly to free us from the then prevalent notion that genome structure is optimized by phenotypic selection. But the persistence of the moniker "selfish DNA" has become an impediment to further understanding of the origin, historical contribution, and contemporary role of transposons in chromosome structure.

Transposons may be an inevitable by-product of the evolution of sequence-specific endonucleases. Complete transposons have been shown to arise from a single cleavage site and an endonuclease gene . Although the successful constitution of a transposon from the recognition sequences used in Ig gene rearrangement and the RAG1 and RAG2 proteins was interpreted as evidence that the V(D)J recombination system evolved from an ancient mobile DNA element, the fact is that the critical components of a transposon and a site-specific rearrangement system are the same (Hiom *et al.*, 1998). Thus questions about the origin of certain kinds of transposons may devolve to questions about the association of sequence-specific DNA binding domains with endonuclease domains.

Although the majority of methylated sequences in a genome can be transposable elements, the view that DNA methylation evolved to control transposons seems implausible in the light of evidence that duplications of any kind trigger methylation in organisms that methylate DNA (Yoder *et al.*, 1997; Garrick *et al.*, 1998; Selker, 1999). And organisms that do not methylate DNA also have mechanisms for detecting duplications and sequestering repeats (Pirrotta, 1997; Sherman and Pillus, 1997; Henikoff, 1998). Genome expansion by duplication is predicated on preventing illegitimate recombination between duplicated sequences.

Although different eukaryotic lineages appear to have invented different mechanisms, what is common to repeat-induced silencing in all eukaryotes is the stable packaging of DNA into "repressive" chromatin. It may be that the evolution of mechanisms that recognize, mark, and sequester duplications into repressive chromatin structures, among which some involve DNA methylation, were the prerequisites for expansion of genomes by endoreduplication at all scales. The additional benefit of such "repressive" mechanisms in minimizing spurious transcription could be secondary sequelae. Because sequence duplication is inherent in transposition, the ability to recognize and repress duplications would serve to minimize both the activity and the adverse impacts of transposons; rendering them genetically invisible and favoring their gradual accumulation.

TRANSPOSONS AND GENOME EVOLUTION IN PLANTS

Although it is known today that transposons comprise a significant fraction of the genomes of many organisms, they eluded discovery through the first half century of genetic analysis and even once discovered, their ubiquity and abundance were not recognized for some time. This genetic invisibility of transposons focuses attention on the mechanisms that control not only transposition, but illegitimate recombination. The thesis is developed that the mechanisms that control transposition are a reflection of the more general capacity of eukaryotic organisms to detect, mark, and retain duplicated DNA through repressive chromatin structures.

The 50 years that have elapsed since the publication of Stebbins' *Variation and Evolution in Plants* have seen extraordinary changes in our understanding of how genomes are structured and how they change in evolution. The book's publication date roughly coincides with the first reports by Barbara McClintock that there are genetic elements capable of transposing to different chromosomal locations in maize plants. The book contains a brief mention of Marcus Rhoades' observation that a standard recessive *a*1 allele of a gene in the anthocyanin biosynthetic pathway can become unstable and revert at a high frequency to the dominant *A*1 allele in a background containing a dominant *Dt* ("dotted") allele (Stebbins Jr., 1950). But transposable elements were not yet common fare, nor was it known that *Dt* is a transposon.

Today we know that transposons constitute a large fraction—even a majority—of the DNA in some species of plants and animals, among them mice, humans, and such agriculturally important plants as corn and wheat. Given what we now know about genome organization, it is paradoxical that the discovery of transposable elements lagged so far behind the discovery of the basic laws of genetic transmission.

And it is equally curious that even when they were discovered, acceptance of their generality and recognition of their ubiquity came so slowly. It is perhaps an understatement to say that McClintock's early communications describing transposition were not widely hailed for their explanatory power. Indeed, McClintock commented in the introduction to her collected papers that the response to her first effort in 1950 to communicate her discovery of transposition in "... a journal with wide readership...," specifically the *Proceedings of the National Academy of Sciences*, convinced her that "... the presented thesis, and evidence for it, could not be accepted by the majority of geneticists or by other biologists" . By contrast, the explanatory power of Watson and Crick's 1953 *Nature* paper on the structure and mode of replication of nucleic acids was recognized immediately (Watson and Crick, 1953).

An informative parallel is provided by the contrast between the immediate recognition of the importance of Darwin's theory of evolution and the long delay between Mendel's articulation of the laws of heredity and their wide acceptance in evolutionary thinking (Carlson, 1966). It can be speculated that this was because Darwin's theory provided immediate explanations in the realm of the perceptible, whereas the hereditary mechanisms under- lying variation were obscure. Variation, in Darwin's view, was continuous. Geneticists sharing his view formed the "biometrical school," devoted to the statistical analysis of inheritance. It was not at all clear how the simple rules derived by Mendel for the hereditary behavior of "differentiating characters" bore on the problem of evolution (Carlson, 1966). The relevance of discontinuous variation or the production of "sports," as morphological mutations were called, was even less obvious, because the biometric approach treated offspring as statistical combinations of parental traits. Thus the idea that the study of mutations was central to understanding evolution was close to unimaginable a century ago.

Equally unimaginable at mid-20th century was the idea that transposable elements are essential to understanding chromosome structure and evolution, much less organismal evolution. The efforts of Bateson and other geneticists had firmly established Mendelian "laws" as the central paradigm of genetics and the identification and mapping of genetic "loci" through the study of mutant alleles was proceeding apace. Because genetic mapping is predicated on the invariance of recombination frequencies, there was plentiful evidence that genes have fixed chromosomal locations. Written at this time, Stebbins' book in general and in particular his third chapter, titled "The Basis of Individual Variation," clearly acknowledges the existence of many chromosomal differences among organisms in a population, including duplications, inversions, translocations, and deletions. At the same time, the book reflects the prevailing view that these "... are not the materials that selection uses to fashion the diverse kinds of organisms which are the products of evolution" . Instead, Stebbins concludes that the majority of evolutionarily important changes in physiology and morphology are attributable to classical genetic "point" mutations.

Another half century has elapsed and the geneticist's "black box," sprung open, spills nucleotide sequences at an ever accelerating pace. Our computers sift through genomes in search of genes, knee-deep in transposons. How could we not have seen them before? The answer is as straightforward as it is mysterious and worthy of consideration: they are invisible to the geneticist. Well, almost invisible. And of course it depends on the geneticist.

THE DISCOVERY OF TRANSPOSITION

The study of unstable mutations that cause variegation dates back to De Vries, who formulated the concept of "ever-sporting varieties" and eventually came to the conclusion that these types of mutations do not obey Mendel's rules . The first person to make substantial sense of their inheritance was the maize geneticist Emerson, who analyzed a variegating allele of the maize *P* locus during the first decades of this century (Emerson, 1914, 1917, 1929). His first paper on the subject opens with the statement that variegation "... is distinguished from other color patterns by its incorrigible irregularity" (Emerson, 1914). What follows is a brilliant analysis of "freak ears" containing large sectors in which the unstable *P* allele has either further mutated or reverted. Emerson was able to capture the behavior of unstable mutations in the Mendelian paradigm by postulating that variegation commenced with the temporary association of some type of inhibitor with a locus required for pigmentation. Emerson's suggestion was that normal pigmentation was restored upon loss of the inhibitor.

Several prominent geneticists, among them Correns and Goldschmidt, dismissed unstable mutations as a special category of "diseased genes" (Goldschmidt, 1938; Fedoroff, 1998). It was their view that little could be learned from the study of such mutations that was relevant to the study of conventional genes. But the drosophilist Demerec and the maize geneticist Rhoades shared Emerson's view that there was no difference in principle between stable and unstable mutations. Indeed, the Rhoades mutation cited in Stebbin's volume illustrates the important point that instability is conditional. Rhoades' experiments had revealed that a standard recessive allele of the maize *A1* locus, isolated decades earlier and in wide use as a stable null allele, could become unstable in a different genetic background. The key ingredient of the destabilizing background was the presence of the *Dt* locus, which caused reversion of the *a1* allele to wild type both somatically and germinally.

In the late 1930s, McClintock had begun to work with broken chromosomes and by the early 1940s she had devised a method for producing deletion mutations commencing with parental plants, each of which contributed a broken chromosome 9 lacking a terminal segment and the telomere. Searching for mutants in the progeny of such crosses, she observed a high frequency of variegating mutants of all kinds . She noted that although reports of the appearance of new mutable genes were relatively rare in the maize literature, she already had isolated 14 new cases of such instability and observed more. She chose to follow the behavior of a locus, which she called *Dissociation* (*Ds*), for its propensity to cause the dissociation of the short arm of chromosome 9 at a position close to the centromere, although she soon appreciated that chromosome dissociation required the presence of another unlinked locus, which she designated the *Activator* (*Ac*) locus (McClintock, 1946, 1947).

By 1948, she had gained sufficient confidence that the *Ds* locus moves to report: "It is now known that the *Ds* locus may change its position in the chromosome" (McClintock, 1948; Fedoroff, 1998). The relationship between the chromosome-breaking *Ds* locus and variegation emerged as McClintock analyzed the progeny of a new variegating mutation of the *C* locus required for kernel pigmentation. She carried out an extraordinary series of painstakingly detailed cytological and genetic experiments on this new mutation, *c-m1*, whose instability was conditional and depended on the presence of the *Ac* locus (McClintock, 1948, 1949; Fedoroff, 1998).

She showed that the origin of the unstable mutation coincided with the transposition of *Ds* from its original position near the centromere on chromosome 9 to a new site at the *C* locus and that when it reverted to a stable wild-type allele, *Ds* disappeared from the locus. Having established that *Ds* could transpose into and out of the *C* locus germinally, she inferred that somatic variegation reflects the frequent transposition of *Ds* during development. Transposition explained both Emerson's and Rhoades' earlier observations. McClintock and Rhoades were good friends, of course, and it is evident from their correspondence that McClintock immediately saw the parallels between the behavior of the *c-m1* mutation and Rhoades' *a1* mutation (Lee Kass, personal communication). The *Ac* and *Ds* elements are transposition-competent and transposition- defective members of a single transposon family. In the ensuing years, McClintock identified and studied a second transposon family, called *Suppressor-mutator* (*Spm*) (McClintock, 1951, 1954).

Her studies on these element families were purely genetic, and she was able to make extraordinary progress in understanding the transposition mechanism because she studied the interactions between a single transposition-competent element and one or a small number of genes with insertions of cognate transposition-defective elements . Two points about this early history of transposition merit emphasis. First, the active elements were denumerable and manageable as genetic entities, despite their propensity to move. Second, the number of different transposon families and family members uncovered genetically was (and still is) small. Hence the genetic impact of transposable elements was limited.

McClintock recognized that the high frequency of new variegating mutations in her cultures was linked to the genetic perturbations associated with the presence of broken chromosomes (McClintock, 1946, 1978). Her inference, extraordinarily prescient, was that transposons are regular inhabitants of the genome, but genetically silent.

PLANT TRANSPOSONS IN THE AGE OF GENOMICS

With the cloning of the maize transposons, first the *Ac* element in my laboratory and later the cognate *En* and *Spm* elements in Heinz Saedler's and my laboratories, the picture began to change (Fedoroff *et al.*, 1983; Pereira *et al.*, 1985; Masson *et al.*, 1987). To begin with, it became obvious immediately that the maize genome contains more copies of a given trans- poson than there are genetically identifiable elements. Although most of these sequences are not complete transposons, there are nonetheless more complete transposons than can be perceived genetically (Fedoroff *et al.*, 1984). Importantly, it was clear almost immediately that a genetically active transposon could be distinguished from one that was genetically silent by its methylation pattern (Fedoroff *et al.*, 1984; Banks and Fedoroff, 1989). Both of these observations bear on the genetic visibility of transposons.

As maize genes and genome segments began to be cloned and sequenced, the discovery of new transposons accelerated. Although the transposons that McClintock identified and studied were DNA transposons, both *gypsy*-like and *copia*-like retrotransposons were soon identified in the maize genome and subsequently in many other plant genomes (Shepherd *et al.*, 1984; Flavell, 1992; Purugganan and Wessler, 1994; White *et al.*, 1994; Suoniemi *et al.*, 1997, 1998). It has also become evident that non-long terminal repeat (LTR) retrotransposons are abundant in maize, as well as other plant genomes (Schwarz-Sommer *et al.*, 1987; Noma *et al.*, 1999). Many additional maize transposon families have been identified through their sequence organization and their presence in or near genes (Spell *et al.*, 1988; Bureau and Wessler, 1992, 1994a, b; Wessler *et al.*, 1995; Bureau *et al.*, 1996). We now know that transposons and retrotransposons comprise half or more of the maize genome (Bennetzen *et al.*, 1998).

What Do Transposons Do?

Commencing with McClintock's elegant analyses of transposon-associated chromosomal rearrangements and extending into the literature of today, the range of transposon-associated genetic changes has continued to expand . Insertion of plant transposons, like almost all known transposons, is accompanied by the duplication of a short flanking sequence of a few base pairs (Schwarz-Sommer *et al.,* 1985a). Plant transposons excise imprecisely, generally leaving part of the duplication at the former insertion site (Schwarz-Sommer *et al.,* 1985a). The consequences of insertion and excision of a transposon therefore depend on the location within the coding sequence and excision of an insertion from an exon commonly results in either an altered gene product or a frame-shift mutation. Transposon insertions can alter transcription and transcript processing, and there are cases in which transposons are processed out of transcripts by virtue of the presence of splice donor and acceptor sequences (Kim *et al.,* 1987; Wessler, 1989; Giroux *et al.,* 1994). Transposons also can promote the movement of large segments of DNA either by transposition or by illegitimate recombination (Courage-Tebbe *et al.,* 1983; Schwartz *et al.,* 1998).

THE PARADOX

One might think that given their abundance, transposable elements would rapidly randomize genome order. Yet the results of a decade of comparative plant genome studies has revealed that gene order is surprisingly conserved between species. Close relationships among genomes have been demonstrated in crop plants belonging to the *Solanaceae,* and the *Graminae,* between *Brassica* crops and *Arabidopsis,* among several legumes, and others (Tanksley *et al.,* 1992; Gale and Devos, 1998; Lager- crantz, 1998). The synteny among the genomes of economically important cereal grasses is so extensive that they are now represented by concentric circular maps (Gale and Devos, 1998). There are rearrangements, but a relatively small number of major inversions and transpositions is required to harmonize the present day maps. Such maps, of course, are crude representations of the genome, and rearrangements can emerge as the level of resolution increases (Tanksley *et al.,* 1988, 1992; Tikhonov *et al.,* 1999). The frequency of rearrangements also can differ markedly and there is evidence that rearrangements are more prevalent just after polyploidization (Song *et al.,* 1995; Gale and Devos, 1998). Even within a conservative lineage, however, some gene families are more heterogeneous in composition and map distribution than others (Leister *et al.,* 1998).

SYNTENY AND DIVERGENCE

What are the useful generalizations? First, synteny can extend down to a very fine level, but it is far from perfect. A detailed sequence comparison of the small region around the maize and sorghum *Adh1* loci reveals a surprising amount of change in a constant framework (Tikhonov *et al.*, 1999). The sorghum and maize genomes are 750 and 2,500 Mbp, respectively. The *Adh1* gene sequences are highly conserved, and complete sequencing revealed that there were seven and 10 additional genes in the homologous regions of maize and sorghum, respectively. The region of homology extends over about 65 kb of the sorghum genome, but occupies more than 200 kb in the maize genome. The gene order and orientation are conserved, although three of the genes found in the sorghum *Adh1* region are not in the maize *Adh1* region.

The genes are located elsewhere in the maize genome, suggesting that they transposed away from the *Adh1* region (Tikhonov *et al.*, 1999). Although homology is confined largely to genes, there are also homologous intergenic regions. There are simple sequence repeats and small transposons, called MITES as a group, scattered throughout this region in both sorghum and maize. MITES are found primarily between genes, but several are in introns. The small MITE transposons are found neither in exons nor in retrotransposons. There are three non-LTR retrotransposons in the maize *Adh1* region and none in the sorghum *Adh1* region (Tikhonov *et al.*, 1999).

The major difference between the maize and sorghum *Adh* regions is the presence of very large continuous blocks of retrotransposons in maize that are not present in sorghum. Although most blocks are between genes, one appears to be inside a gene sequence. They are present in many, but not all intergenic regions. There is a relatively long stretch of almost 40 kb containing four genes in maize and seven genes in sorghum, which contains no retrotransposon blocks in maize and in which there is about 10 kb of extensive homology, some genic and some intergenic. Thus synteny extends down to a relatively fine level and includes both genic and intergenic sequences.

4

Plant Cell Structure

Plants are unique among the eukaryotes, organisms whose cells have membrane-enclosed nuclei and organelles, because they can manufacture their own food. Chlorophyll, which gives plants their green colour, enables them to use sunlight to convert water and carbon dioxide into sugars and carbohydrates, chemicals the cell uses for fuel. Like the fungi, another kingdom of eukaryotes, plant cells have retained the protective cell wall structure of their prokaryotic ancestors. The basic plant cell shares a similar construction motif with the typical eukaryote cell, but does not have centrioles, lysosomes, intermediate filaments, cilia, or flagella, as does the animal cell. Plant cells do, however, have a number of other specialized structures, including a rigid cell wall, central vacuole, plasmodesmata, and chloroplasts. Although plants (and their typical cells) are non-motile, some species produce gametes that do exhibit flagella and are, therefore, able to move about.

Plants can be broadly categorized into two basic types: vascular and nonvascular. Vascular plants are considered to be more advanced than nonvascular plants because they have evolved specialized tissues, namely xylem, which is involved in structural support and water conduction, and phloem, which functions in food conduction. Consequently, they also possess roots, stems, and leaves, representing a higher form of organization that is characteristically absent in plants lacking vascular tissues. The nonvascular plants, members of the division Bryophyta, are usually no more than an inch or two in height because they do not have adequate support, which is provided by vascular tissues to other plants, to grow bigger. They also are more dependent on the environment that surrounds them to maintain appropriate amounts of moisture and, therefore, tend to inhabit damp, shady areas.

It is estimated that there are at least 260,000 species of plants in the world today. They range in size and complexity from small, nonvascular mosses to giant sequoia trees, the largest living organisms, growing as tall as 330 feet (100 meters). Only a tiny percentage of those species are directly used by people for food, shelter, fiber, and medicine. Nonetheless, plants are the basis for the Earth's ecosystem and food web, and without them complex animal life forms (such as humans) could never have evolved. Indeed, all living organisms are dependent either directly or indirectly on the energy produced by photosynthesis, and the byproduct of this process, oxygen, is essential to animals. Plants also reduce the amount of carbon dioxide present in the atmosphere, hinder soil erosion, and influence water levels and quality.

Plants exhibit life cycles that involve alternating generations of diploid forms, which contain paired chromosome sets in their cell nuclei, and haploid forms, which only possess a single set. Generally these two forms of a plant are very dissimilar in appearance. In higher plants, the diploid generation, the members of which are known as sporophytes due to their ability to produce spores, is usually dominant and more recognizable than the haploid gametophyte generation. In Bryophytes, however, the gametophyte form is dominant and physiologically necessary to the sporophyte form.

Animals are required to consume protein in order to obtain nitrogen, but plants are able to utilize inorganic forms of the element and, therefore, do not need an outside source of protein. Plants do, however, usually require significant amounts of water, which is needed for the photosynthetic process, to maintain cell structure and facilitate growth, and as a means of bringing nutrients to plant cells. The amount of nutrients needed by plant species varies significantly, but nine elements are generally considered to be necessary in relatively large amounts. Termed macroelements, these nutrients include calcium, carbon, hydrogen, magnesium, nitrogen, oxygen, phosphorus, potassium, and sulfur. Seven microelements, which are required by plants in smaller quantities, have also been identified: boron, chlorine, copper, iron, manganese, molybdenum, and zinc.

Thought to have evolved from the green algae, plants have been around since the early Paleozoic era, more than 500 million years ago. The earliest fossil evidence of land plants dates to the Ordovician Period (505 to 438 million years ago). By the Carboniferous Period, about 355 million years ago, most of the Earth was covered by forests of primitive vascular plants, such as lycopods (scale trees) and gymnosperms (pine trees, ginkgos).

Angiosperms, the flowering plants, didn't develop until the end of the Cretaceous Period, about 65 million years ago—just as the dinosaurs became extinct.

- Cell Wall-Like their prokaryotic ancestors, plant cells have a rigid wall surrounding the plasma membrane. It is a far more complex structure, however, and serves a variety of functions, from protecting the cell to regulating the life cycle of the plant organism.

- Chloroplasts-The most important characteristic of plants is their ability to photosynthesize, in effect, to make their own food by converting light energy into chemical energy. This process is carried out in specialized organelles called chloroplasts.
- Endoplasmic Reticulum-The endoplasmic reticulum is a network of sacs that manufactures, processes, and transports chemical compounds for use inside and outside of the cell. It is connected to the double-layered nuclear envelope, providing a pipeline between the nucleus and the cytoplasm. In plants, the endoplasmic reticulum also connects between cells via the plasmodesmata.
- Golgi Apparatus-The Golgi apparatus is the distribution and shipping department for the cell's chemical products. It modifies proteins and fats built in the endoplasmic reticulum and prepares them for export as outside of the cell.
- Microfilaments-Microfilaments are solid rods made of globular proteins called actin. These filaments are primarily structural in function and are an important component of the cytoskeleton.
- Microtubules-These straight, hollow cylinders are found throughout the cytoplasm of all eukaryotic cells (prokaryotes don't have them) and carry out a variety of functions, ranging from transport to structural support.
- Mitochondria-Mitochondria are oblong shaped organelles found in the cytoplasm of all eukaryotic cells. In plant cells, they break down carbohydrate and sugar molecules to provide energy, particularly when light isn't available for the chloroplasts to produce energy.
- Nucleus-The nucleus is a highly specialized organelle that serves as the information processing and administrative centre of the cell. This organelle has two major functions: it stores the cell's hereditary material, or DNA, and it coordinates the cell's activities, which include growth, intermediary metabolism, protein synthesis, and reproduction (cell division).
- Peroxisomes-Microbodies are a diverse group of organelles that are found in the cytoplasm, roughly spherical and bound by a single membrane. There are several types of microbodies but peroxisomes are the most common.
- Plasmodesmata-Plasmodesmata are small tubes that connect plant cells to each other, providing living bridges between cells.
- Plasma Membrane-All living cells have a plasma membrane that encloses their contents. In prokaryotes and plants, the membrane is the inner layer of protection surrounded by a rigid cell wall. These

membranes also regulate the passage of molecules in and out of the cells.

- Ribosomes-All living cells contain ribosomes, tiny organelles composed of approximately 60 percent RNA and 40 percent protein. In eukaryotes, ribosomes are made of four strands of RNA. In prokaryotes, they consist of three strands of RNA.
- Vacuole-Each plant cell has a large, single vacuole that stores compounds, helps in plant growth, and plays an important structural role for the plant.

Leaf Tissue Organization-The plant body is divided into several organs: roots, stems, and leaves. The leaves are the primary photosynthetic organs of plants, serving as key sites where energy from light is converted into chemical energy. Similar to the other organs of a plant, a leaf is comprised of three basic tissue systems, including the dermal, vascular, and ground tissue systems. These three motifs are continuous throughout an entire plant, but their properties vary significantly based upon the organ type in which they are located.

HISTORY OF PLANT CELL

The term *cell,* as first used by Robert Hooke in 1665 signified an apparently empty space or lumen, surrounded by walls. We now know, of course, that the space is far from empty, and that rigid cell walls as seen by Hooke in thin slices of cork, are not ubiquitous in multicellular organisms. Indeed, the wall became to be regarded as the definitive structure of the cell, and when in the 1830s, the zoologist Schwann was able to recognise structures in cartilage resembling plant parenchymatous cell walls, the concept of the cell as the basic biological unit common to all organisms was born. Definitions have changed considerably in the subsequent century and a half, and, in particular, the cell wall is now seen in its proper perspective as being a structure, albeit of great importance, but restricted to plants and existing only outside the true cell. Nevertheless, the general concept of the cell as the basic minimlum unit of life remains.

Since all organisms need to perform a number of essential functions merely in order to survive, both as individuals and as species, it should not be surprising to find a basic unity between the cells of all organisms. Each cell, at least in the early stages of its development, possesses the capacity to synthesize complex substances from simple ones, to liberate and transform the potential chemical energy of highly reduced compounds, to react to internal and external stimuli, to control the influx and efflux of materials across the limiting cell membranes and to regulate its activities in relation to the information contained in its individual store, or stores, of hereditary genetic material. Evolution has solved the problems posed by these requirements in more or less identical ways in all organisms, and thus the basic processes, activities, and structures of each individual plant cell are similar, not only to other plant cells, but also to all other eucaryotic cells. This book concentrates on the unifying features of plant cells and relates them to present knowledge and general theories of molecular biology. It should not be forgotten, however, that cells are characterised as much by their diversity as their unity. A wide range of different cell types with varying specialized functions are necessary for the life of the higher green plant; however, the origin of cell heterogeneity is a topic outside the scope of this present book.

The basic structures of an undifferentiated plant cell. The cell proper is delimited by the *plasma membrane* (or *plasmalemma*) which is of unit membrane construction. Outside the plasma membrane, and thus actually *extra-cellular,* is the *cell wall.* The cell wall is normally closely appressed to the plasma membrane and in meristematic cells is thin and relatively weak. During differentiation various specialized wall structures develop; depending on the function of the mature cell, the walls may become relatively massive and extremely strong through the deposition of rigid, highly cross-linked polymeric substances.

Adjacent protoplasts (i.e. the cells proper) are connected across the cell walls by narrow cytoplasmic channels, bounded by the plasma membrane, known as *plasmodesmata* (PD). Within the cell a number of separate compartments, and interconnecting compartments, delimited by membranes, may be recognised. *Vacuoles* (V) are prominent, apparently empty spaces, spherical and numerous in the meristematic cell but irregular, very large, and coalescent in the mature expanded cell. Vacuoles serve as intracellular dust-bins—repositories for unwanted and often toxic byproducts of metabolism—and may also have functions similar to the *lysosomes* of animal cells. They are bounded by a single membrane known as the *tonoplast.* The nucleus (N), a major compartment in most cells, comprises *a nuclear envelope* possessing many large *nuclear pores* (NP) and *nucleoplasm,* the ground substance in which the hereditary material, *chromatin,* and the *nucleolus* (NU) lie. The nucleus is the principal site of the hereditary material of the cell, although both plastids and mitochondria also contain DNA. The material outside the nuclear envelope is commonly known as *cytoplasm.*

Ramifying throughout the cytoplasm, and occasionally connected to the outer membrane of the nuclear envelope, the cisternae of the *endoplasmic reticulum* act to integrate the biosynthetic functions of the cell. The endoplasmic reticulum is generally classified into two types: *rough endoplasmic reticulum*

(RER), which has *ribosomes* attached to its outer face; and *smooth endoplasmic reticulum* (SER) which is not involved in protein synthesis. The endoplasmic reticulum may also, on occasion, be seen to be associated with stacks of vesicles (VE) known collectively as *dictyosomes* (D) or *Golgi bodies.* The endoplasmic reticulum and the dictyosomes are responsible for the formation and secretion of cellular membranes.

Three other membrane-bound compartments remain, each concerned with an aspect of energy or intermediary metabolism. *Plastids* (P), undifferentiated in meristematic cells and present only as *proplastids,* represent a general class of organelle in which the *chloroplast* is the characteristic member. *Mitochondria* (M) are smaller, but also bounded by a double membrane, and similarly involved in energy metabolism. As mentioned above, both mitochondria and plastids contain their own stores of hereditary material.

The final compartments, in contrast, are bound by only a single membrane and do not contain hereditary material; these are known as *microbodies* (MB) and often contain dense, granular, or even crystalline contents.

Within the cytoplasm just inside the plasma membrane lie long narrow cylinders known as microtubules (MT); microtubules function in a number of processes in which orientation of cellular components is important. Finally, plant cells contain many fine fibrils, known as *microfilaments,* which appear to be contractile in function and to be composed of a material similar to actin, one of the contractile components of muscle. The 'typical' plant cell does not exist, of course, and the meristematic cell has only been chosen since it possesses all the essential characteristics of plant cells. Many of the cellular components are only present in very simple forms in meristematic cells.

PIONEERING PATHFINDERS

Darwin's Dilemma

Like so many aspects of natural science, the beginnings of the search for life's earliest history date from the mid-1800s and the writings of Charles Darwin (1809–1882), who in *On the Origin of Species* first focused attention on the missing Precambrian fossil record and the problem it posed to his theory of evolution: "There is another... difficulty, which is much more serious. I allude to the manner in which species belonging to several of the main divisions of the animal kingdom suddenly appear in the lowest known [Cambrian-age] fossiliferous rocks... If the theory be true, it is indisputable that before the lowest Cambrian stratum was deposited, long periods elapsed... and that during these vast periods, the world swarmed with living creatures... [But] to the question why we do not find rich fossiliferous deposits belonging to these assumed earliest periods before the Cambrian system, I can give no satisfactory answer. The case at present must remain inexplicable; and may be truly urged as a valid argument against the views here entertained".

Darwin's dilemma begged for solution. And although this problem was to remain unsolved—the case "inexplicable"—for more than 100 years, the intervening century was not without bold pronouncements, dashed dreams, and more than little acid acrimony.

J. W. Dawson and the "Dawn Animal of Canada"

Among the first to take up the challenge of Darwin's theory and its most vexing problem, the missing early fossil record, was John William Dawson (1820–1899), Principal of McGill University and a giant in the history of North American geology. Schooled chiefly in Edinburgh, Scotland, the son of strict Scottish Presbyterians, Dawson was a staunch Calvinist and devout antievolutionist (O'Brien, 1971).

In 1858, a year before publication of Darwin's opus, specimens of distinctively green- and white-layered limestone collected along the Ottawa River to the west of Montreal were brought to the attention of William E. Logan, Director of the Geological Survey of Canada. Because the samples were known to be ancient (from "Laurentian" strata, now dated at about 1,100 million years) and exhibited layering that Logan supposed too regular to be purely inorganic, he displayed them as possible "pre-Cambrian fossils" at various scientific conferences, where they elicited spirited discussion but gained little acceptance as remnants of early life. In 1864, however, Logan brought specimens to Dawson who not only confirmed their biologic origin but identified them as fossilized shells of giant foraminiferans, huge oversized versions of tiny calcareous protozoal tests. So convinced was Dawson of their biologic origin that a year later, in 1865, he formally named the putative fossils *Eozoon canadense*, the "dawn animal of Canada."

Dawson's interpretation was questioned almost immediately (King and Rowney, 1866), the opening shot of a fractious debate that raged on until 1894 when specimens of *Eozoon* were found near Mt. Vesuvius and shown to be geologically young ejected blocks of limestone, their "fossil-like" appearance the result of inorganic alteration and veining by the green metamorphic mineral serpentine (O'Brien, 1970).

Yet despite the overwhelming evidence, Dawson continued to press his case for the rest of his life, spurred by his deeply held belief that discovery of his "dawn animal" had exposed the greatest missing link in the entire fossil record, a gap so enormous that it served to unmask the myth of evolution's claimed continuity and left Biblical creation as the only answer: "There is no link whatever in geological fact to connect Eozoon with the Mollusks, Radiates, or Crustaceans of the succeeding [rock record]... these stand before us as distinct creations. [A] gap... yawns in our imperfect geological record.

Of actual facts [with which to fill this gap], therefore, we have none; and those evolutionists who have regarded the dawn-animal as an evidence in their favour, have been obliged to have recourse to supposition and assumption" (Dawson, 1875, p. 227). (In part, Dawson was right. In the fourth and all later editions of *The Origin*, Darwin cited the great age and primitive protozoal relations of *Eozoon* as consistent with his theory of evolution, just the sort of "supposition and assumption" that Dawson found so distressing.)

Founder of Precambrian Paleobiology

Fortunately, Dawson's debacle would ultimately prove to be little more than a distracting detour on the path to progress, a redirection spurred initially by the prescient contributions of the American paleontologist Charles Doolittle Walcott (1850–1927).

Like Dawson before him, Walcott was enormously energetic and highly influential (Yochelson, 1967, 1997). He spent most of his adult life in Washington, DC, where he served as the CEO of powerful scientific organizations—first, as Director of the U.S. Geological Survey (1894–1907), then Secretary of the Smithsonian Institution (1907–1927) and President of the National Academy of Sciences (1917–1923).

Surprisingly, however, Walcott had little formal education. As a youth in northern New York State he received but 10 years of schooling, first in public schools and, later, at Utica Academy (from which he did not graduate). He never attended college and had no formally earned advanced degrees (a deficiency more than made up for in later life when he was awarded honorary doctorates by a dozen academic institutions).

In 1878, as a 28-year-old apprentice to James Hall, Chief Geologist of the state of New York and acknowledged dean of American paleontology, Walcott was first introduced to stromatolites—wavy layered moundshaped rock masses laid down by ancient communities of mat-building microbes—Cambrian-age structures near the town of Saratoga in eastern New York State.

Named *Cryptozoon* (meaning "hidden life"), these cabbagelike structures would in later years form the basis of Walcott's side of a nasty argument known as the "*Cryptozoon* controversy."

A year later, in July, 1879, Walcott was appointed to the newly formed U.S. Geological Survey. Over the next several field seasons, he and his comrades charted the geology of sizable segments of Arizona, Utah, and Nevada, including unexplored parts of the Grand Canyon, where in 1883 he first reported discovery of Precambrian specimens of *Cryptozoon* (Walcott, 1883).

Other finds soon followed, with the most startling in 1899—small, millimeter-sized black coaly discs that Walcott named *Chuaria* and interpreted to be "the remains of... compressed conical shell[s]," possibly of primitive brachiopods (Walcott, 1899). Although *Chuaria* is now known to be a large single-celled alga, rather than a shelly invertebrate,

Walcott's specimens were indeed authentic fossils, the first true cellularly preserved Precambrian organisms ever recorded.

After the turn of the century, Walcott moved his field work northward along the spine of the Rocky Mountains, focusing first in the Lewis Range of northwestern Montana, from which he reported diverse stromatolitelike structures (Walcott, 1906) and, later, chains of minute celllike bodies he identified as fossil bacteria (Walcott, 1915).

His studies in the Canadian Rockies, from 1907 to 1925, were even more rewarding, resulting in discovery of an amazingly well-preserved assemblage of Cambrian algae and marine invertebrates—the famous Burgess Shale Fauna that to this day remains among the finest and most complete samples of Cambrian life known to science (Walcott, 1911; Gould, 1989).

Walcott's contributions are legendary—he was the first discoverer in Precambrian rocks of *Cryptozoon* stromatolites, of cellularly preserved algal plankton (*Chuaria*), and of possible fossil bacteria, all capped by his pioneering investigations of the benchmark Burgess Shale fossils. The acknowledged founder of Precambrian paleobiology (Schopf, 1970), Walcott was first to show, nearly a century ago and contrary to accepted wisdom, that a substantial fossil record of Precambrian life actually exists.

A. C. SEWARD AND THE *CRYPTOZOON* CONTROVERSY

The rising tide in the development of the field brought on by Walcott's discoveries was not yet ready to give way to a flood. Precambrian fossils continued to be regarded as suspect, a view no doubt bolstered by Dawson's *Eozoon* debacle but justified almost as easily by the scrappy nature of the available evidence. Foremost among the critics was Albert Charles Seward (1863–1941), Professor of Botany at the University of Cambridge and the most widely known and influential paleobotanist of his generation.

Because practically all claimed Precambrian fossils fell within the purview of paleobotany—whether supposed to be algal, like *Cryptozoon* stromatolites, or even bacterial—Seward's opinion had special impact.

In 1931, in *Plant Life Through the Ages*, the paleobotanical text used worldwide, Seward assessed the "algal" (that is, cyanobacterial) origin of *Cryptozoon* as follows: "The general belief among American geologists and several European authors in the organic origin of *Cryptozoon* is... not justified by the facts. [Cyanobacteria] or similar primitive algae may have flourished in Pre-Cambrian seas and inland lakes; but to regard these hypothetical plants as the creators of reefs of *Cryptozoon* and allied structures is to make a demand upon imagination inconsistent with Wordsworth's definition of that quality as 'reason in its most exalted mood'" (Seward, 1931, pp. 86–87).

Seward was even more categorical in his rejection of Walcott's report of fossil bacteria: "It is claimed that sections of a Pre-Cambrian limestone from Montana show minute bodies similar in form and size to cells and cell-chains of existing [bacteria].... These and similar contributions... are by no means convincing.... We can hardly expect to find in Pre- Cambrian rocks any actual proof of the existence of bacteria... " .

Seward's 1931 assessment of the science was mostly on the mark. Mistakes had been made. Mineralic, purely inorganic objects had been misinterpreted as fossil. Better and more evidence, carefully gathered and dispassionately considered, was much needed.

But his dismissive rejection of *Cryptozoon* and his bold assertion that "we can hardly expect to find in Pre-Cambrian rocks any actual proof of the existence of bacteria" turned out to be misguided. Yet his influence was pervasive. It took another 30 years and a bit of serendipity to put the field back on track.

EMERGENCE OF A NEW FIELD OF SCIENCE

In the mid-1960s—a full century after Darwin broached the problem of the missing early fossil record—the hunt for early life began to stir, and in the following two decades the flood-gates would finally swing wide open. But this surge, too, had harbingers, now dating from the 1950s.

A Benchmark Discovery by an Unsung Hero

The worker who above all others set the course for modern studies of ancient life was Stanley A. Tyler (1906–1963) of the University of Wisconsin, the geologist who in 1953 discovered the now famous mid-Precambrian (2,100-million-year-old) microbial assemblage petrified in carbonaceous cherts of the Gunflint Formation of Ontario, Canada. A year later, together with Harvard paleobotanist Elso S. Barghoorn (1915–1984), Tyler published a short note announcing the discovery (Tyler and Barghoorn, 1954), a rather sketchy report that on the basis of study of petrographic thin sections documents that the fossils are indigenous to the deposit but fails to note either the exact provenance of the find or that the fossils are present within, and were actually the microbial builders of, large *Cryptozoon*like stromatolites (an association that, once recognized, would prove key to the development of the field).

Substantive, full-fledged reports would come later—although not until after Tyler's untimely death, an event that cheated him from receiving the great credit he deserved—but this initial short article on "the oldest structurally preserved organisms that clearly exhibit cellular differentiation and original carbon complexes which have yet been discovered in pre-Cambrian sediments" (Tyler and Barghoorn, 1954) was a benchmark, a monumental "first."

Contributions of Soviet Science

At about the same time, in the mid-1950s, a series of articles by Boris Vasil'evich Timofeev (1916–1982) and his colleagues at the Institute of Precambrian Geochronology in Leningrad (St. Petersburg) reported discovery of microscopic fossil spores in Precambrian siltstones of the Soviet Union.

In thin sections, like those studied by Tyler and Barghoorn, fossils are detected within the rock, entombed in the mineral matrix, so the possibility of laboratory contamination can be ruled out.

But preparation of thin sections requires special equipment, and their microscopic study is tedious and time-consuming. A faster technique, pioneered for Precambrian studies in Timofeev's lab, is to dissolve a rock in mineral acid and concentrate the organic-walled microfossils in the resulting sludgelike residue.

This maceration technique, however, is notoriously subject to error-causing contamination—and because during these years there was as yet no established early fossil record with which to compare new finds, mistakes were easy to make. Although Timofeev's laboratory was not immune, much of his work has since proved sound (Schopf, 1992), and the technique he pioneered to ferret-out microfossils in Precambrian shaley rocks is now in use worldwide.

Famous Figures Enter the Field

Early in the 1960s, the fledgling field was joined by two geologic heavyweights, an American, Preston Cloud (1912–1991), and an Australian, Martin Glaessner (1906–1989), both attracted by questions posed by the abrupt appearance and explosive evolution of shelly invertebrate animals that marks the start of the Phanerozoic Eon. A feisty leader in the development of Precambrian paleobiology, Cloud was full of energy, ideas, opinions, and good hard work.

His Precambrian interests were first evident in the late-1940s, when he argued in print that although the known Early Cambrian fossil record is woefully incomplete it is the court of last resort and, ultimately, the only court that matters (Cloud, 1948).

By the 1960s, he had become active in the field, authoring a paper that to many certified the authenticity of the Tyler- Barghoorn Gunflint microfossils

(Cloud, 1965) and, later, a series of reports adding new knowledge of the early microbial fossil record (Cloud and Licari, 1968; Cloud *et al.*, 1975; Cloud and Morrison, 1980). But above all, he was a gifted synthesist, showing his mettle in a masterful article of 1972 that set the stage for modern understanding of the interrelated atmospheric- geologic-biologic history of the Precambrian planet (Cloud, 1972).

In the early 1960s, a second prime player entered this now fast-unfolding field, Martin Glaessner (1906–1989), of the University of Adelaide in South Australia. A scholarly, courtly, old-school professor, Glaessner was the first to make major inroads toward understanding the (very latest) Precambrian record of multicelled animal life (Radhakrishna, 1991; McGowran, 1994).

In 1947, three years before Glaessner joined the faculty at Adelaide, Reginald C. Sprigg announced his discovery of fossils of primitive softbodied animals, chiefly imprints of saucer-sized jellyfish, at Ediacara, South Australia (Sprigg, 1947).

Although Sprigg thought the fossil-bearing beds were Cambrian in age, Glaessner showed them to be Precambrian (albeit marginally so), making the Ediacaran fossils the oldest animals known.

Together with his colleague, Mary Wade, Glaessner spent much of the rest of his life working on this benchmark fauna, bringing it to international attention in the early 1960s (Glaessner, 1962) and, later, in a splendid monograph (Glaessner, 1984).

With Glaessner in the fold, the stage was set. Like a small jazz band—Tyler and Barghoorn trumpeting microfossils in cherts, Timofeev beating on fossils in siltstones, Cloud strumming the early environment, Glaessner the earliest animals—great music was about to be played. At long last, the curtain was to rise on the missing record of Precambrian life.

Breakthrough to the Present

My own involvement dates from 1960, when as a sophomore in college I became enamored with the problem of the missing Precambrian fossil record, an interest that was to become firmly rooted during the following few years, when I was the first of Barghoorn's graduate students to focus on early life.

I have recently recounted in some detail my recollections of those heady days (Schopf, 1999) and need not reiterate the story here. Suffice it to note that virtually nothing had been published on the now-famous Gunflint fossils in the nearly 10 years that had passed between the Tyler-Barghoorn 1954 announcement of the find and my entry into graduate school in June, 1963.

Then, quite unexpectedly, in October of that year, Stanley Tyler passed away at the age of 57, never to see the ripened fruits of his long-term labor reach the published page. Within a year thereafter, a series of events that would shape the field began to unfold, set off first by a squabble between Barghoorn and Cloud as to who would scoop whom in a battle for credit over the Gunflint fossils (Schopf, 1999).

By late 1964 this spat had been settled, with Cloud electing to hold off publication of his paper "illustrating some conspicuous Gunflint nannofossils and discussing their implications until Barghoorn could complete his part of a descriptive paper with the by-then deceased Tyler" (Cloud, 1983, p. 23).

The two articles appeared in *Science* in 1965, first Barghoorn and Tyler's "Microorganisms from the Gunflint chert" (Barghoorn and Tyler, 1965), followed a few weeks later by Cloud's contribution, "Significance of the Gunflint (Precambrian) microflora" (Cloud, 1965). Landmark papers they were!

Unlike the 1954 Tyler-Barghoorn announcement of discovery of the Gunflint fossils, which had gone largely unnoticed, the Barghoorn-Tyler 1965 article—backed by Cloud's affirmation of its significance—generated enormous interest. Yet it soon became apparent that acceptance of ancient life would come only grudgingly.

The well had been poisoned by Dawson's debacle, the *Cryptozoon* controversy, Seward's criticism—object lessons that had been handed down from professor to student, generation to generation, to become part of accepted academic lore. Moreover, it was all too obvious that the Gunflint organisms stood alone.

Marooned in the remote Precambrian, they were isolated by nearly a billion and half years from all other fossils known to science, a gap in the known fossil record nearly three times longer than the entire previously documented history of life. Skepticism abounded. Conventional wisdom was not to be easily dissuaded.

The question was asked repeatedly: "Couldn't this whole business be some sort of fluke, some hugely embarrassing awful mistake?"

As luck would have it, the doubts soon could be laid to rest. During field work the previous year (and stemming from a chance conversation with a local oil company geologist by the name of Helmut Wopfner), Barghoorn had collected a few hand-sized specimens of Precambrian stromatolitic black chert in the vicinity of Alice Springs, deep in the Australian outback. Once the Gunflint paper had been completed, I was assigned to work on the samples, which quite fortunately contained a remarkable cache of new microscopic fossils, most nearly indistinguishable from extant cyanobacteria and almost all decidedly better preserved than the Gunflint microbes.

Although the age of the deposit (the Bitter Springs Formation) was known only approximately, it seemed likely to be about 1,000 million years, roughly half as old as the Gunflint chert. Barghoorn and I soon sent a short report to *Science* (Barghoorn and Schopf, 1965), publication of which—viewed in light of the earlier articles on the Gunflint organisms—not only served to dispell lingering doubts about whether Precambrian fossils might be some sort of fluke, but seemed to show that the early fossil record was surprisingly richer and easier to unearth than anyone had dared imagine. Indeed, it now appears that the only truly odd thing about the Gunflint and Bitter Springs fossils is that similar finds had not been made even earlier.

Walcott had started the train down the right track only for it to be derailed by the conventional wisdom that the early history of life was unknown and evidently unknowable, a view founded on the assumption that the tried and true techniques of the Phanerozoic hunt for large fossils would prove equally rewarding in the Precambrian. Plainly put, this was wrong.

LESSONS FROM THE HUNT

The Gunflint and Bitter Springs articles of 1965 charted a new course, showing for the first time that a search strategy centered on the peculiarities of the Precambrian fossil record would pay off.

The four keys of the strategy, as valid today as they were three decades ago, are to search for (*i*) microscopic fossils in (*ii*) black cherts that are (*iii*) fine-grained and (*iv*) associated with *Cryptozoon*-like structures. Each part plays a role.

(*i*) Megascopic eukaryotes, the large organisms of the Phanerozoic, are now known not to have appeared until shortly before the beginning of the Cambrian—except in immediately sub-Cambrian strata, the hunt for large body fossils in Precambrian rocks was doomed from the outset.

(*ii*) The blackness of a chert commonly gives a good indication of its organic carbon content—like fossil-bearing coal deposits, cherts rich in petrified organic-walled microfossils are usually a deep jet black color.

(*iii*) The fineness of the quartz grains making up a chert provides another hint of its fossil-bearing potential—cherts subjected to the heat and pressure of geologic metamorphism are often composed of recrystallized large grains that give them a sugary appearance whereas cherts that have escaped fossil-destroying processes are made up of cryptocrystalline quartz and have a waxy glasslike luster.

(*iv*) *Cryptozoon*-like structures (stromatolites) are now known to have been produced by flourishing microbial communities, layer upon layer of microscopic organisms that make up localized biocoenoses.

Stromatolites permineralized by fine-grained chert early during diagenesis represent promising hunting grounds for the fossilized remnants of the microorganisms that built them.

Measured by virtually any criterion one might propose studies of Precambrian life have burst forth since the mid-1960s to culminate in recent years in discovery of the oldest fossils known, petrified cellular microbes nearly 3,500 million years old, more than three-quarters the age of the Earth.

Precambrian paleobiology is thriving—the vast majority of all scientists who have ever investigated the early fossil record are alive and working today; new discoveries are being made at an ever quickening clip—progress set in motion by the few bold scientists who blazed this trail in the 1950s and 1960s, just as their course was charted by the Dawsons, Walcotts, and Sewards, the pioneering pathfinders of the field.

And the collective legacy of all who have played a role dates to Darwin and the dilemma of the missing Precambrian fossil record he first posed. After more than a century of trial and error, of search and final discovery, those of us who wonder about life's early history can be thankful that what was once "inexplicable" to Darwin is no longer so to us.

CRYSTALLINE INCLUSIONS WITHIN PLANT CELLS

Many plant cells contain crystalline inclusions of different chemical composition and shape. Crystalline aggregations are called druses, bundles of needle-shaped crystals are termed raphids. Scanning electron microscopic images: Top picture: Calcium oxalate druse in the mesophyll cells of an oleander leaf *(Nerium oleander)*. Typical druse shape of dicots. Middle picture: Calcium oxalate needles (raphids) of a vanilla root (Orchidaceae). Typical raphid bundle of monocots. Lowest picture: Silicate bodies of silicate cells in the epidermis of *Schizachyrium sanguineum* (a gramineaen species of the old world tropics). Characteristic mineralization of a gramineaen cell.

The Cell Wall

Except for very few examples, plant cells are surrounded by a cellulose containing cell wall. It is flexible and distortable during growth, but loses its ability for distortion after growth has stopped, while a limited flexibility remains. Because of these changes, it is distinguished between primary and secondary cell walls. As we will see when talking about electron microscopic pictures of the cell wall, both forms differ mainly in the arrangement of their cellulose microfibrils. While they are unorganized within the amorphous matrix of the primary cell wall, they are organized into several ordered layers that are arranged one on top of the other at right angles in the secondary cell wall. The secondary cell walls of many cells, especially those of vascular tissues, are incrusted with strengthening material. Two important ones are:

- lignin, the ground substance of wood and
- suberin, the ground substance of cork

Their details are reviewed here. In addition, secondary walls contain often phenolic oxidation products that lends them a dark colour (red to black with various shades).

Tissues

The branch of science that is committed to the study of tissues is called histology. Though with plants the term plant anatomy is heard just as often. The term tissue stems originally from a misinterpretation. N. GREW coined it in 1682 since he thought that the scaffolding of the cell wall would consist of very fine threads and that the organization of the plant would resemble the layering of a larger number of Brussels lace on top of one another. He spoke in this context of cell tissue (*contexus cellulosus*). Despite this misapprehension the term was used from now on and was even adopted by animal histologists. From our contemporary point of view tissues are combinations of cells, organs are functional unities of an organism. The basic organs of the phanerogamous plant are leaf, stem and root.

Organs consist of different tissues, the leaves for example of dermal tissue, assimilation tissue and vascular tissues. A tissue again may have differently structured cell types. The vascular tissue contains thus the cells of the xylem and those of the phloem. Based upon a suggestion of C. W.v. NÄGELI it is distinguished between local regions of cell division, the meristems, and permanent tissues. Meristems are characterized by cell divisions, while this is an exceptional feature in permanent tissues. The latter are as a rule differentiated and often specialized. Depending on their functional properties can they be grouped into the following categories:

dermal tissues (epidermis, cork, bark)

ground tissues (parenchyma)

assimilation tissues (palisade parenchyma, spongy mesophyll)

collenchyma and sclerenchyma tissues (collenchyma, sclereids, fibers)

vascular tissues (vessel elements, xylem, phloem)

A. de BARY published his basic work on histology : "Vergleichende Anatomie der Vegetationsorgane der Phanerogamen und Farne" (Comparative Anatomy of the Vegetation Organs of Phanerogames and Ferns) in 1877. It contains the following statement:

" The elements of every tissue are derived from the cells of the meristems, every element has thus originally all the properties of a cell.

During their differentiation the main difference produced is that one group keeps these properties throughout their lives, while the other loses them. The first retains the ability of independent growth and is thus able to divide.

As a result, these cells can themselves become meristems. The other group loses the ability to divide and to grow independently with continuing differentiation. These cells stop usually growing at all. In some cases a continuing real enlargement of such elements takes place as a result of their feeding by neighboring cells."

The details of the differentiation process will be discussed elsewhere. Here, I will present some examples of recurring division patterns. They allow some conclusions about shape and size of the cells.

Probably the most important principle of vegetable differentiation is the polarity that is developed during embryogenesis (ontogenesis) at the first division of the fertilized egg. It determines the main axis of the vegetation body irreversibly. After that, shoot and root develop independently. This polarity is also known as root-shoot-polarity. Cell divisions that take place in a plane that is perpendicular to the surface of the next surface of the organ are called anticlinal. Those that take place in parallel to it are called periclinal. Often the cells divide unequally and as the result are two cells of different sizes. The rule of thumb says that the smaller one remains in the already existing physiological state, while the bigger one differentiates or specializes into a certain direction. Exceptions exist. In guard cell development, for example, the smaller cell differentiates stronger than the bigger one.

One of the most striking features of unequally differentiated cells is their uneven enlargement. Some cells divide without noticeable increase in volume, while others stop dividing and grow considerably. Vegetable growth is thus caused mainly by an increase in the volume of single cells. Since the enlargement of every cell is restricted by the resistance of neighboring ones and since the cells are cemented to each other rather strongly by the middle lamina, considerable tensions are inevitable.

A cell that enlarges evenly into all directions results in a sphere. Neighboring cells have different sizes due to both asynchrony of cell division and different individual ontogenesis, so that the walls between neighboring cells differ in size, too. As a result, the shape of a cell that enlarges evenly into all directions looks more like a many-sided body, a so-called isodiametric cell or polyhedral, than like a sphere.

Elongation of cells is often directed. It occurs usually in parallel to the axis of the organ in question. The resulting cell shape is that of a spindle. It is also called prosenchymatous. If all cells of a certain area are elongating (like those of the apical meristems of roots and shoots), then the respective organ elongates evenly, while the growth of cells at only one side of the organ will lead to bending. We will discuss this process further when discussing the directed growth of plants.

The mentioned tensions and altered plant shapes, that are caused by uneven cell divisions and differences in volume enlargement of single cells show that the cormus of the plant cannot be described by simple geometric relations.

The irregularities of cell divisions (different division activities of single cells or cell groups, unequal divisions, etc.), the differences in the ability to elongate and the way of specialization are therefore the causes of the histological variability and the formation of a specific shape of the plant (morphogenesis). Although every step itself seems like a deviation of the normal, all steps are controlled by the plant's genome. Growth and differentiation are thus exactly matched and co-ordinated in a way that gives the developing plant the form that is specific for the respective species.

There is accordingly a flow of information between the cells of an organism that helps regulating their activity. This is most impressive in the development of symmetric shapes at both cellular and organ level. But how is the control of growth and the co-ordination of differentiation achieved?

In 1965, J. BONNER of the California Institute of Technology at Pasadena presented a model that explains how genetically determined co-ordination points for the development of specialized cells develop from a single cell. It may seem rather hypothetical, but it is a useful working hypothesis that prompts to search for the demanded regulators.

EUKARYOTIC CELLS AND PLANT CELL

Plant cells are eukaryotic cells that differ in several key respects from the cells of other eukaryotic organisms. Their distinctive features include:

- A large central vacuole, a water-filled volume enclosed by a membrane known as the *tonoplast* maintains the cell's turgor, controls movement of molecules between the cytosol and sap, stores useful material and digests waste proteins and organelles.
- A cell wall composed of cellulose and hemicellulose, pectin and in many cases lignin, is secreted by the protoplast on the outside of the cell membrane. This contrasts with the cell walls of fungi (which are made of chitin), and of bacteria, which are made of peptidoglycan.
- Specialised cell–cell communication pathways known as plasmodesmata, pores in the primary cell wall through which the plasmalemma and endoplasmic reticulum of adjacent cells are continuous.

- Plastids, the most notable being the chloroplasts, which contain chlorophyll a green coloured pigment which is used for absorbing sunlight and is used by a plant to make its own food in the process is known as photosynthesis. Other types of plastid are the amyloplasts, specialized for starch storage, elaioplasts specialized for fat storage, and chromoplasts specialized for synthesis and storage of pigments. As in mitochondria, which have a genome encoding 37 genes, plastids have their own genomes of about 100–120 unique genes and, it is presumed, arose as prokaryotic endosymbionts living in the cells of an early eukaryotic ancestor of the land plants and algae.
- Cell division by construction of a phragmoplast as a template for building a cell plate late in cytokinesis is characteristic of land plants and a few groups of algae, notably the Charophytes and the Order Trentepohliales
- The sperm of bryophytes and pteridophytes have flagellae similar to those in animals, but higher plants, (including Gymnosperms and flowering plants) lack the flagellae and centrioles that are present in animal cells.

Cell Types

- Parenchyma cells are living cells that have really diverse functions ranging from storage and support to photosynthesis and phloem loading (transfer cells). Apart from the xylem and phloem in their vascular bundles, leaves are composed mainly of parenchyma cells. Some parenchyma cells, as in the epidermis, are specialized for light penetration and focusing or regulation of gas exchange, but others are among the least specialized cells in plant tissue, and may remain totipotent, capable of dividing to produce new populations of undifferentiated cells, throughout their lives. Parenchyma cells have thin, permeable primary walls enabling the transport of small molecules between them, and their cytoplasm is responsible for a wide range of biochemical functions such as nectar secretion, or the manufacture of secondary products that discourage herbivory. Parenchyma cells that contain many chloroplasts and are concerned primarily with photosynthesis are called chlorenchyma cells. Others, such as the majority of the parenchyma cells in potato tubers and the seed cotyledons of legumes, have a storage function.
- Collenchyma cells – collenchyma cells are alive at maturity and have only a primary wall. These cells mature from meristem derivatives that initially resemble parenchyma, but differences quickly become apparent. Plastids do not develop, and the secretory apparatus (ER

and Golgi) proliferates to secrete additional primary wall. The wall is most commonly thickest at the corners, where three or more cells come in contact, and thinnest where only two cells come in contact, though other arrangements of the wall thickening are possible.

Pectin and hemicellulose are the dominant constituents of collenchyma cell walls of dicotyledon angiosperms, which may contain as little as 20% of cellulose in *Petasites*. Collenchyma cells are typically quite elongated, and may divide transversely to give a septate appearance.

The role of this cell type is to support the plant in axes still growing in length, and to confer flexibility and tensile strength on tissues. The primary wall lacks lignin that would make it tough and rigid, so this cell type provides what could be called plastic support – support that can hold a young stem or petiole into the air, but in cells that can be stretched as the cells around them elongate. Stretchable support (without elastic snap-back) is a good way to describe what collenchyma does. Parts of the strings in celery are collenchyma.

- Sclerenchyma cells – Sclerenchyma cells (from the Greek skleros, *hard*) are hard and tough cells with a function in mechanical support. They are of two broad types – sclereids or stone cells and fibres. The cells develop an extensive secondary cell wall that is laid down on the inside of the primary cell wall. The secondary wall is impregnated with lignin, making it hard and impermeable to water. Thus, these cells cannot survive for long' as they cannot exchange sufficient material to have active metabolism. Sclerenchyma cells are typically dead at functional maturity, and the cytoplasm is missing, leaving an empty central cavity.

Functions for sclereid cells (hard cells that give leaves or fruits a gritty texture) include discouraging herbivory, by damaging digestive passages in small insect larval stages, and physical protection (a solid tissue of hard sclereid cells form the pit wall in a peach and many other fruits). Functions of fibres include provision of load-bearing support and tensile strength to the leaves and stems of herbaceous plants. Sclerenchyma fibres are not involved in conduction, either of water and nutrients (as in the xylem) or of carbon compounds (as in the phloem), but it is likely that they may have evolved as modifications of xylem and phloem initials in early land plants.

Tissue Types

The major classes of cells differentiate from undifferentiated meristematic cells (analogous to the stem cells of animals) to form the tissue structures of roots, stems, leaves, flowers, and reproductive structures. Xylem cells are elongated cells with lignified secondary thickening of the cell walls. Xylem cells are specialised for conduction of water, and first appeared in plants during their transition to land in the Silurian period more than 425 million years ago. The possession of xylem defines the vascular plants or Tracheophytes. Xylem tracheids are pointed, elongated xylem cells, the simplest of which have continuous primary cell walls and lignified secondary wall thickenings in the form of rings, hoops, or reticulate networks. More complex tracheids with valve-like perforations called bordered pits characterise the gymnosperms. The ferns and other pteridophytes and the gymnosperms have only xylem tracheids, while the angiosperms also have xylem vessels. Vessel members are hollow xylem cells aligned end-to-end, without end walls that are assembled into long continuous tubes. The bryophytes lack true xylem cells, but their sporophytes have a water-conducting tissue known as the hydrome that is composed of elongated cells of simpler construction.

Phloem is a specialised tissue for food conduction in higher plants. The conduction of food is a complex process that is carried in the plant with the help of special cell called phloem cells. These cells conduct inter-and intra-cellular fluid (food – proteins and other essential elements required by the plant for its metabolism) through the process of osmosis. This phenomenon is called ascent of sap in plants. Phloem consists of two cell types, the sieve tubes and the intimately-associated companion cells. The sieve tube elements lack nuclei and ribosomes, and their metabolism and functions are regulated by the adjacent nucleate companion cells. Sieve tubes are joined end-to-end with perforate end-plates between known as *sieve plates*, which allow transport of photosynthate between the sieve elements. The companion cells, connected to the sieve tubes via plasmodesmata, are responsible for loading the phloem with sugars. The bryophytes lack phloem, but moss sporophytes have a simpler tissue with analogous function known as the leptome.

Plant epidermal cells are specialised parenchyma cells covering the external surfaces of leaves, stems and roots. The epidermal cells of aerial organs arise from the superficial layer of cells known as the *tunica* (L1 and L2 layers) that covers the plant shoot apex, whereas the cortex and vascular tissues arise from innermost layer of the shoot apex known as the *corpus* (L3 layer). The epidermis of roots originates from the layer of cells immediately beneath the root cap.

The epidermis of all aerial organs, but not roots, is covered with a cuticle made of the polyester cutin with a superficial layer of waxes. The epidermal cells of the primary shoot are thought to be the only plant cells with the biochemical capacity to synthesize cutin. Several cell types may be present in the epidermis. Notable among these are the stomatal guard cells, glandular and clothing hairs or trichomes, and the root hairs of primary roots. In the shoot epidermis of most plants, only with really good chloroplasts.

ANATOMY OF PLANT CELL

The cell is the basic unit of life. Plant cells (unlike animal cells) are surrounded by a thick, rigid cell wall. The following is a glossary of plant cell anatomy terms. Amyloplast-an organelle in some plant cells that stores starch. Amyloplasts are found in starchy plants like tubers and fruits.

ATP-ATP is short for adenosine triphosphate; it is a high-energy molecule used for energy storage by organisms. In plant cells, ATP is produced in the cristae of mitochondria and chloroplasts. Cell membrane-the thin layer of protein and fat that surrounds the cell, but is inside the cell wall. The cell membrane is semipermeable, allowing some substances to pass into the cell and blocking others. Cell wall-a thick, rigid membrane that surrounds a plant cell. This layer of cellulose fiber gives the cell most of its support and structure. The cell wall also bonds with other cell walls to form the structure of the plant.

Centrosome-(also called the "microtubule organizing centre") a small body located near the nucleus-it has a dense centre and radiating tubules. The centrosomes is where microtubules are made. During cell division (mitosis), the centrosome divides and the two parts move to opposite sides of the dividing cell. Unlike the centrosomes in animal cells, plant cell centrosomes do not have centrioles.

Chlorophyll-chlorophyll is a molecule that can use light energy from sunlight to turn water and carbon dioxide gas into sugar and oxygen (this process is called photosynthesis). Chlorophyll is magnesium based and is usually green.

Chloroplast-an elongated or disc-shaped organelle containing chlorophyll. Photosynthesis (in which energy from sunlight is converted into chemical energy-food) takes place in the chloroplasts.

Christae-(singular crista) the multiply-folded inner membrane of a cell's mitochondrion that are finger-like projections. The walls of the cristae are the site of the cell's energy production (it is where ATP is generated).

Cytoplasm-the jellylike material outside the cell nucleus in which the organelles are located.

Golgi body-(also called the golgi apparatus or golgi complex) a flattened, layered, sac-like organelle that looks like a stack of pancakes and is located near the nucleus. The golgi body packages proteins and carbohydrates into membrane-bound vesicles for "export" from the cell.

Granum-(plural grana) A stack of thylakoid disks within the chloroplast is called a granum.

Mitochondrion-spherical to rod-shaped organelles with a double membrane. The inner membrane is infolded many times, forming a series of projections (called cristae). The mitochondrion converts the energy stored in glucose into ATP (adenosine triphosphate) for the cell. Nuclear membrane-the membrane that surrounds the nucleus. Nucleolus-an organelle within the nucleus-it is where ribosomal RNA is produced.

Nucleus-spherical body containing many organelles, including the nucleolus. The nucleus controls many of the functions of the cell (by controlling protein synthesis) and contains DNA (in chromosomes). The nucleus is surrounded by the nuclear membrane

Photosynthesis-a process in which plants convert sunlight, water, and carbon dioxide into food energy (sugars and starches), oxygen and water. Chlorophyll or closely-related pigments (substances that colour the plant) are essential to the photosynthetic process.

Ribosome-small organelles composed of RNA-rich cytoplasmic granules that are sites of protein synthesis.

Rough endoplasmic reticulum-(rough ER) a vast system of interconnected, membranous, infolded and convoluted sacks that are located in the cell's cytoplasm (the ER is continuous with the outer nuclear membrane). Rough ER is covered with ribosomes that give it a rough appearance. Rough ER transport materials through the cell and produces proteins in sacks called cisternae (which are sent to the Golgi body, or inserted into the cell membrane).

Smooth endoplasmic reticulum-(smooth ER) a vast system of interconnected, membranous, infolded and convoluted tubes that are located in the cell's cytoplasm (the ER is continuous with the outer nuclear membrane). The space within the ER is called the ER lumen. Smooth ER transport materials through the cell. It contains enzymes and produces and digests lipids (fats) and membrane proteins; smooth ER buds off from rough ER, moving the newly-made proteins and lipids to the Golgi body and membranes

Stroma-part of the chloroplasts in plant cells, located within the inner membrane of chloroplasts, between the grana.

Thylakoid disk-thylakoid disks are disk-shaped membrane structures in chloroplasts that contain chlorophyll. Chloroplasts are made up of stacks of thylakoid disks; a stack of thylakoid disks is called a granum. Photosynthesis (the production of ATP molecules from sunlight) takes place on thylakoid disks.

Vacuole-a large, membrane-bound space within a plant cell that is filled with fluid. Most plant cells have a single vacuole that takes up much of the cell. It helps maintain the shape of the cell.

EARLY EVOLUTION AND THE ORIGIN OF CELLS

Darwin noticed the sudden appearance of several major animal groups in the oldest known fossiliferous rocks. "If [my] theory be true, it is indisputable that before the lowest Cambrian stratum was deposited... the world swarmed with living creatures," he wrote, noting that he has "no satisfactory answer" to the "question why we do not find rich fossiliferous deposits belonging to these assumed earliest periods" (Darwin, 1859). In "Solution to Darwin's Dilemma: Discovery of the Missing Precambrian Record of Life", J. William Schopf points out that, one century later, one decade after the publication of Stebbins' *Variation and Evolution in Plants* , the situation had not changed. The known history of life extended only to the beginning of the Cambrian Period, some 550 million years ago. This state of affairs would soon change, notably due to three papers published in *Science* in 1965 by E.S. Barghoorn and S.A. Tyler (1965), Preston Cloud (1965), and E.S. Barghoorn and J.W. Schopf (1965). Schopf tells of the predecessors who anticipated or made possible the work reported in the three papers, and of his own and others' contributions to current knowledge, which places the oldest fossils known, in the form of petrified cellular microbes, nearly 3,500 million years ago, seven times older than the Cambrian and reaching into the first quarter of the age of the Earth.

Lynn Margulis, M.F. Dolan, and R. Guerrero set their thesis right in the title of their contribution: "The Chimeric Eukaryote: Origin of the Nucleus from the Karyomastigont in Amitochondriate Protists". The karyomastigont is an organellar system composed at least of a nucleus with protein connectors to one (or more) kinetosome.

The ancestral eukaryote cell was a chimera between a thermoacidophilic archaebacterium and a heterotrophic eubacterium, a "bacterial consortium" that evolved into a heterotrophic cell, lacking mitochondria at first. Cells with free nuclei evolved from karyomastigont ancestors at least five times, one of them becoming the mitochondriate aerobic ancestor of most eukaryotes. These authors aver that only two major categories of organisms exist: prokaryotes and eukaryotes. The Archaea, making a third category according to Carl Woese and others (Woese *et al.,* 1990), should be considered bacteria and classified with them.

The issue of shared genetic organelle origins is also a subject, if indirect, of the paper by Jeffrey D. Palmer and colleagues ("Dynamic Evolution of Plant Mitochondrial Genomes: Mobile Genes and Introns, and Highly Variable Mutation Rates,). The mitochondrial DNA (mtDNA) of flowering plants (angiosperms) can be more than 100 times larger than is typical of animals. Plant mitochondrial genomes evolve rapidly in size, both by growing and shrinking; within the cucumber family, for example, mtDNA varies more than six fold. Palmer and collaborators have investigated more than 200 angiosperm species and uncovered enormous pattern heterogeneities, some lineage specific.

The authors reveal numerous losses of mt ribosomal protein genes (but only rarely of respiratory genes), virtually all in some lineages; yet, most ribosomal protein genes have been retained in other lineages. High rates of functional transfer of mt ribosomal protein genes to the nucleus account for many of the loses. The authors show that plant mt genomes can increase in size, acquiring DNA sequences by horizontal transfer. Their striking example is a group I intron in the mt *cox* gene, an invasive mobile element that may have transferred between species more than 1,000 independent times during angiosperm evolution. It has been known for more than a decade that the rate of nucleotide substitution in angiosperm mtDNA is very low, 50–100 times lower than in vertebrate mtDNA. Palmer *et al.* have now discovered fast substitution rates in *Pelargonium* and *Plantago*, two distantly related angiosperms.

AN APPRECIATION

Ledyard Stebbins, born January 6, 1906, was deeply fond of nature all his life, starting with his experiences around Seal Harbor, Maine, at about four years old. When he was still young, his mother contracted tuberculosis, and the family moved west to try to find a more healthy climate for her—first to Pasadena, then to Colorado Springs. An important formative period of Ledyard's life was spent at Cate School, in Santa Barbara, where he studied for four years. During those years, he was, by his own account, shy and relatively unpopular; but he learned to ride horseback, explored the Santa Ynez Mountains, and fell under the influence of the botanist Ralph Hoffmann, who taught him much about the plants and natural history of that lovely area.

Enrolling in Harvard University in 1924, Ledyard at first had difficulty defining his major, but the summer between his freshman and sophomore years was spent investigating the plants around Bar Harbor, Maine, the family home, and brought him into contact with Edgar T. Wherry, professor of botany from the University of Pennsylvania and a specialist in ferns, who encouraged his botanical interests. When he enrolled for the fall semester of 1925 at Harvard, he had decided to pursue a botanical career. But during his time at Harvard, his love of classical music, which was to be an important element for the remainder of his life, was awakened and nourished, as he participated in music classes and choruses, and was encouraged by some powerful and encouraging faculty members and students. Continuing on in the Harvard Graduate School, Ledyard was caught in the cross-fire between those, like Merrit Lyndon Fernald, who took a classical view of botany and plant classification, and the more modern approaches of Karl Sax, who was applying cytogenetic principles to developing a deeper understanding of plants. Thanks to judicious efforts by Paul Mangelsdorf and others, his dissertation was finally approved; but it was a struggle; he graduated in 1931.

One of the key events in Ledyard's early career was his attending the International Botanical Congress at Cambridge, England, in 1930; there he met Edgar Anderson, who was to become a lifetime friend and colleague; Irene Manton; and C.D. Darlington, whose classical "Recent Advances of Cytology," was still in the future. These and other contacts greatly encouraged his interest in and enthusiasm for botany and botanists, which was to be sustained for the rest of his life.

After he obtained his Ph.D., Ledyard Stebbins spent the years 1931–1935 at Colgate University, which he described years later as unhappy years, but it is not clear why this was the case. With an associate, Professor Percy Sanders, he undertook the cytogenetic study of Paeonia, which was the first of a series of essentially biosystematic investigations of diverse plant groups that were to characterize the remainder of his research career. During this time, he discovered complex structural heterozygosity in the western North American species of the genus, an exciting find that was to fuel his enthusiasm for further cytogenetic investigations.

In 1935, Professor Ernest Brown Babcock of the University of California, Berkeley, offered Stebbins a research position in connection with his investigations of the genus Crepis, which he accepted with alacrity. Met at the train station by his fellow Harvard student Rimo Bacigalupi, he plunged into this project with enthusiasm. Also at Berkeley, he began his lifetime preoccupation with Democratic politics, working actively in the 1936 Roosevelt election, and from there onward. After four years on Professor Babcock's grant, Stebbins was appointed to the faculty at Berkeley, and began to teach a course in the principles of evolution, which helped him to generalize his thoughts and finally to his preparing the classical work, "Variation and Evolution in Plants," whose 50th anniversary we are celebrating in this symposium.

In his research efforts, Ledyard began investigating the American species of Crepis (most species of the genus are Eurasian, but there are some very interesting offshoots in North America), outlining the evolutionary features of this group as a pillar complex of polyploids, with the base chromosome number 2n = 22, but widespread polyploids, characteristically apomictic, linking their more narrowly-distributed diploid progenitors. He also began investigating grasses, first Bromus and then

Triticeae, with the objective of developing perennial grasses that would provide forage on the dry rangelands of California, and which eventually led to his extensive studies of the genus Dactylis, which he pursued throughout its native range in western Eurasia and North Africa in the decades to follow. Never successful, this quest nonetheless led Stebbins to many interesting discoveries, and broadened the scope of his knowledge of the details of evolution in plants in such a way as to expand the coverage of and insights provided in his landmark book.

In the early 1940s, Stebbins began working actively with Carl Epling on the genetics of Linanthus parryae, an annual of the Mohave Desert in which the prevalence of white or blue flowers in individual populations was held at the time to have resulted from random drift. He also started an active association with Theodosius Dobzhansky, centering around Dobzhansky's efforts at Mather; he regarded Epling, Dobzhansky, and Edgar Anderson as his closest and most influential professional associates.

In 1947, Ledyard Stebbins spent three months at Columbia University in New York, delivering the Jesup Lectures; and these lectures, expanded and elaborated, became *Variation and Evolution in Plants,* the most important book on plant evolution of the 20th century. I first met him in 1950, on a Sierra Club outing, and he was as encouraging to me at the age of 14 as I could have imagined. It seemed to me later that his own rather unhappy and lonely childhood led him naturally to an appreciation for young people, a lifetime interest in connection with which he made significant contributions to the lives of many young scholars. I maintained a strong friendship with him for the remaining half-century of his life.

The first period of Ledyard Stebbins' botanical life extended from 1925, when his serious interest in plants was kindled at Harvard, to 1935, when he arrived at Berkeley; the second, highly productive period, from there to 1950, when "Variation and Evolution in Plants" was published. In that same year, he answered an invitation from the University to establish a department of genetics at the Davis campus, and entered the third period of his professional life. And my, how he loved Davis, its growth, its variety, and its accessibility to all. He was proud of his work at Davis, proud of the growing campus as it matured, pleased with his own contributions, and always contented living there. In 1971, after Dobzhansky's retirement from Rockefeller University, he was influential in recruiting both Dobzhansky and his associate Francisco Ayala, to Davis, where they made outstanding contributions. With retirement, he traveled widely, for example, teaching in Chile during the time of the 1973 coup, and visiting Australia, Africa, Europe, and other parts of the world in teaching, visiting with his colleagues, and, as always, enjoying students.

SOLUTION TO DARWIN'S DILEMMA

In 1859, in On the Origin of Species, Darwin broached what he regarded to be the most vexing problem facing his theory of evolution—the lack of a rich fossil record predating the rise of shelly invertebrates that marks the beginning of the Cambrian Period of geologic time (H"550 million years ago), an "inexplicable" absence that could be "truly urged as a valid argument" against his all embracing synthesis. For more than 100 years, the "missing Precambrian history of life" stood out as one of the greatest unsolved mysteries in natural science. But in recent decades, understanding of life's history has changed markedly as the documented fossil record has been extended sevenfold to some 3,500 million years ago, an age more than three-quarters that of the planet itself. This long-sought solution to Darwin's dilemma was set in motion by a small vanguard of workers who blazed the trail in the 1950s and 1960s, just as their course was charted by a few pioneering pathfinders of the previous century, a history of bold pronouncements, dashed dreams, search, and final discovery.

In 1950, when Ledyard Stebbins' *Variation and Evolution in Plants* first appeared, the known history of life—the familiar progression from spore-producing to seed-producing to flowering plants, from marine invertebrates to fish, amphibians, then reptiles, birds, and mammals—extended only to the beginning of the Cambrian Period of the Phanerozoic Eon, roughly 550 million years ago. Now, after a half-century of discoveries, life's history looks strikingly different—an immense early fossil record, unknown and assumed unknowable, has been uncovered to reveal an evolutionary progression dominated by microbes that stretches seven times farther into the geologic past than previously was known. This essay is an abbreviated history of how and by whom the known antiquity of life has been steadily extended, and of lessons learned in this still ongoing hunt for life's beginnings.

5

Plant Reproduction

The plant kingdom is divided into many classes and divisions. In 1883, Eichler separated the plant kingdom (Plantae) into two subkingdoms, the first being Cryptogamae (bacteria, algae, fungi, molds, liverworts, moss and ferns); the second being Phanerogamae (including gymnosperms and angiosperms). There have been many changes to the classification system, as we learn more about the genetics of different species. More recent taxonomy created new kingdoms that separated fungi, algae, and molds from what we commonly consider as plants. Plantae was separated into two divisions: Bryophytes (non-vascular) and Tracheophytes (vascular). Of the vascular plants, angiosperms (the flowering plants) make up approximately 300,000 of the 312,000 known number of living species. The number of 312,000 is probably a gross under-estimation. For anyone interested there is an estimated 20,000 species of non-vascular plants.

There are many different methods of reproduction that plants have developed over the centuries. If a plant's method of reproduction makes it easier to cross species or hybridized, it is more likely that its offspring will be better adapted to a changing environment For more than 8,000 years, human beings have also been affecting plant development by purposefully breeding new characteristics into plants, creating new species. The most important fact regarding reproduction is that every living organism has, in its cell's nucleus, a set of chromosomes. Chromosomes carry genes that have all the information necessary to make new individual organisms.

REPRODUCTION IN PLANTS

Plant reproduction is the process by which plants generate new individuals, or offspring. Reproduction is either sexual or asexual. Sexual reproduction is the formation of offspring by the fusion of gametes. Asexual reproduction is the formation of offspring without the fusion of gametes.

Sexual reproduction results in offspring genetically different from the parents. Asexual offspring are genetically identical except for mutation. In higher plants, offspring are packaged in a protective seed, which can be long lived and can disperse the offspring some distance from the parents. In flowering plants (angiosperms), the seed itself is contained inside a fruit, which may protect the developing seeds and aid in their dispersal.

Sexual Reproduction in Angiosperms: Ovule Formation

All plants have a life cycle that consists of two distinct forms that differ in size and the number of chromosomes per cell. In flowering plants, the large, familiar form that consists of roots, shoots, leaves, and reproductive structures (flowers and fruit) is diploid and is called the sporophyte. The sporophyte produces haploid microscopic gametophytes that are dependent on tissues produced by the flower. The reproductive cycle of a flowering plant is the regular, usually seasonal, cycling back and forth from sporophyte to gametophyte.

The flower produces two kinds of gametophytes, male and female. The female gametophyte arises from a cell within the ovule, a small structure within the ovary of the flower. The ovary is a larger structure within the flower that contains and protects usually many ovules. Flowering plants are unique in that their ovules are entirely enclosed in the ovary. The ovary itself is part of a larger structure called the carpel, which consists of the stigma, style, and ovary. Each ovule is attached to ovary tissue by a stalk called the funicle. The point of attachment of the funicle to the ovary is called the placenta.

As the flower develops from a bud, a cell within an ovule called the archespore enlarges to form an embryo-sac mother cell (EMC). The EMC divides by meiosis to produce four megaspores. In this process the number of chromosomes is reduced from two sets in the EMC to one set in the megaspores, making the megaspores haploid. Three of the four megaspores degenerate and disappear, while the fourth divides mitotically three times to produce eight haploid cells. These cells together constitute the female gametophyte, called the embryo sac.

The eight embryo sac cells differentiate into two synergids, three antipodal cells, two fused endosperm nuclei, and an egg cell. The mature embryo sac is situated at the outer opening (micropyle) of the ovule, ready to receive the sperm cells delivered by the male gametophyte.

Pollen

The male gametophyte is the mature pollen grain. Pollen is produced in the anthers, which are attached at the distal end of filaments. The filament and anther together constitute the stamen, the male sex organ. Flowers usually produce many stamens just inside of the petals. As the flower matures, cells in the anther divide mitotically to produce pollen mother cells (PMC). The PMCs divide by meiosis to produce haploid microspores in groups of four called tetrads. The microspores are housed within a single layer of cells called the tapetum, which provides nutrition to the developing pollen grains.

Each microspore develops a hard, opaque outer layer called the exine, which is constructed from a lipoprotein called sporopollenin. The exine has characteristic pores, ridges, or projections that can often be used to identify a species, even in fossil pollen. The microspore divides mitotically once or twice to produce two or three haploid nuclei inside the mature pollen grain. Two of the nuclei function as sperm nuclei that can eventually fuse with the egg and endosperm nuclei of the embryo sac, producing an embryo and endosperm, respectively.

For sexual fusion to take place, however, the pollen grain must be transported to the stigma, which is a receptive platform on the top of the style, an elongated extension on top of the carpel(s). Here the moist surface or chemicals cause the pollen grain to germinate. Germination is the growth of a tube from the surface of a pollen grain. The tube is a sheath of pectin, inside of which is a solution of water, solutes, and the two or three nuclei, which lack any cell walls. Proper growth of the pollen tube requires an aqueous solution of appropriate solute concentration, as well as nutrients such as boron, which may aid in its synthesis of pectin.

At the apex of the tube are active ribosomes and endoplasmic reticulum (types of cell organelles) involved in protein synthesis. Pectinase and a glucanase (both enzymes that break down carbohydrates) probably maintain flexibility of the growing tube and aid in penetration. The pollen tube apex also releases ribonucleic acid (RNA) and ribosomes into the tissues of the style. The tube grows to eventually reach the ovary, where it may travel along intercellular spaces until it reaches a placenta. Through chemical recognition, the pollen tube changes its direction of growth and penetrates through the placenta to the ovule. Here the tube reaches the embryo sac lying close to the micropyle, and sexual fertilization takes place.

Double Fertilization

Fertilization in flowering plants is unique among all known organisms, in that not one but two cells are fertilized, in a process called double fertilization. One sperm nucleus in the pollen tube fuses with the egg cell in the embryo sac, and the other sperm nucleus fuses with the diploid endosperm nucleus. The fertilized egg cell is a zygote that develops into the diploid embryo of the sporophyte. The fertilized endosperm nucleus develops into the triploid endosperm, a nutritive tissue that sustains the embryo and seedling. The only other known plant group exhibiting double fertilization is the Gnetales in the genus *Ephedra,* a nonflowering seed plant. However, in this case the second fertilization product degenerates and does not develop into endosperm.

Double fertilization begins when the pollen tube grows into one of the two synergid cells in the embryo sac, possibly as a result of chemical attraction to calcium. After penetrating the synergid, the apex of the pollen tube breaks open, releasing the two sperm nuclei and other contents into the synergid. As the synergid degenerates, it envelops the egg and endosperm cells, holding the two sperm nuclei close and the other expelled contents of the pollen tube. The egg cell then opens and engulfs the sperm cell, whose membrane breaks apart and allows the nucleus to move near the egg nucleus. The nuclear envelopes then disintegrate, and the two nuclei combine to form the single diploid nucleus of the zygote. The other sperm cell fuses with the two endosperm nuclei, forming a single triploid cell, the primary endosperm cell, which divides mitotically into the endosperm tissue.

Double fertilization and the production of endosperm may have contributed to the great ecological success of flowering plants by accelerating the growth of seedlings and improving survival at this vulnerable stage. Faster seedling development may have given flowering plants the upper hand in competition with gymnosperm seedlings in some habitats, leading to the abundance of flowering plants in most temperate and tropical regions. Gymnosperms nevertheless are still dominant at higher elevations and latitudes, and at low elevations in the Pacific Northwest coniferous forests, such as the coastal redwoods. The reasons for these patterns are still controversial.

The Seed

The seed is the mature, fertilized ovule. After fertilization, the haploid cells of the embryo sac disintegrate. The maternally derived diploid cells of the ovule develop into the hard, water-resistant outer covering of the seed, called the testa, or seed coat. The diploid zygote develops into the embryo, and the triploid endosperm cells multiply and provide nutrition. The testa usually shows a scar called the hilum where the ovule was originally attached to the funicle. In some seeds a ridge along the testa called the raphe shows where the funicle originally was pressed against the ovule. The micropyle of the ovule usually survives as a small pore in the seed coat that allows passage of water during germination of the seed.

In some species, the funicle develops into a larger structure on the seed called an aril, which is often brightly colored, juicy, and contains sugars that are consumed by animals that may also disperse the seed (as in nutmeg, arrowroot, oxalis, and castor bean). This is distinct from the fruit, which forms from the ovary itself. The embryo consists of the cotyledon(s), epicotyl, and hypocotyl. The cotyledons resemble small leaves, and are usually the first photosynthetic organs of the plant. The portion of the embryo above the cotyledons is the epicotyl, and the portion below is the hypocotyl. The epicotyl is an apical meristem that produces the shoot of the growing plant and the first true leaves after germination.

The hypocotyl develops into the root. Often the tip of the hypocotyl, the radicle, is the first indication of germination as it breaks out of the seed. Flowering plants are classified as monocotyledons or dicotyledons (most are now called eudicots) based on the number of cotyledons produced in the embryo. Common monocotyledons include grasses, sedges, lilies, irises, and orchids; common dicotyledons include sunflowers, roses, legumes, snapdragons, and all nonconiferous trees.

The endosperm may be consumed by the embryo, as in many legumes, which use the cotyledons as a food source during germination. In other species the endosperm persists until germination, when it is used as a food reserve. In grains such as corn and wheat, the outer layer of the endosperm consists of thick-walled cells called aleurone, which are high in protein.

The Fruit

The fruit of a flowering plant is the mature ovary. As seeds mature, the surrounding ovary wall forms a protective structure that may aid in dispersal. The surrounding ovary tissue is called the pericarp and consists of three layers. From the outside to inside, these layers are the exocarp, mesocarp, and endocarp. The exocarp is usually tough and skinlike. The mesocarp is often thick, succulent, and sweet. The endocarp, which surrounds the seeds, may be hard and stony, as in most species with fleshy fruit, such as apricots.

A fruit is termed simple if it is produced by a single ripened ovary in a single flower (apples, oranges, apricots). An aggregated fruit is a cluster of mature ovaries produced by a single flower (blackberries, raspberries, strawberries). A multiple fruit is a cluster of many ripened ovaries on separate flowers growing together in the same inflorescence (pineapple, mulberry. A simple fruit may be fleshy or dry. A fleshy simple fruit is classified as a berry (grape, tomato, papaya), pepo (cucumber, watermelon, pumpkin), hesperidium (orange), drupe (apricot), or pome (apple).

Dry simple fruits have a dry pericarp at maturity. They may or may not dehisce, or split, along a seam to release the seeds. A dehiscent dry fruit is classified as legume or pod (pea, bean), silique or silicle (mustard), capsule (poppy, lily), or follicle (milkweed, larkspur, columbine). An indehiscent dry fruit that does not split to release seeds is classified as an achene (sunflower, buttercup, sycamore), grain or caryopsis (grasses such as corn, wheat, rice, barley), schizocarp (carrot, celery, fennel), winged samara (maple, ash, elm), nut (acorn, chestnut, hazelnut), or utricle (duckweed family). Some fruiting bodies contain non-ovary tissue and are sometimes called pseudocarps. The sweet flesh of apples and pears, for example, is composed not of the pericarp but the receptacle, or upper portion, of the flowering shoot to which petals and other floral organs are attached.

Fruiting bodies of all kinds function to protect and disperse the seeds they contain. Protection can be physical (hard coverings) or chemical (repellents of seed predators). Sweet, fleshy fruits are attractive food for birds and mammals that consume seeds along with the fruit and pass the seeds intact in their fecal matter, which can act as a fertilizer. Dry fruits are usually adapted for wind dispersal of seeds, as for example with the assistance of winglike structures or a fluffy pappus that provides buoyancy. The diversity of fruiting bodies reflects in part the diversity of dispersal agents in the environment, which select for different fruit size, shape, and chemistry.

Pollination and Pollinators

Pollination is the movement of pollen from the stamens to the stigma, where germination and growth of the pollen tube occur. Most (approximately 96 percent) of all flowering plant species are hermaphroditic (possess both sexual functions within a plant, usually within every flower), and thus an individual can be pollinated by its own pollen or by pollen from another individual. Seed produced through self-pollination ("selfed" seed) is often inferior in growth, survival, and fecundity to seed produced through outcross pollination ("outcrossed" seed). As a result, in most species there is strong natural selection to maximize the proportion of outcrossed seed (the "outcrossing rate").

Flowering plants are unusual among seed plants in their superlative exploitation of animals (primarily insects) as agents of outcross pollination. The outcross pollination efficiency of insects, birds, and mammals (primarily bats) may have contributed to both the abundance and diversity of flowering plants. Abundance may have increased because of less wastage of energy and resources on unsuccessful pollen and ovules. Diversity may have increased for two reasons. First, insects undoubtedly have selected for a wide variety of floral forms that provide different rewards (pollen and nectar) and are attractive in appearance (colour juxtaposition, size, shape) and scent (sweet, skunky) in different ways to different pollinators. Second, faithfulness of pollinators to particular familiar flowers may have reduced hybridization and speeded evolutionary divergence and the production of new species.

Although flowering plants first appeared after most of the major groups of insects had already evolved, flowering plants probably caused the evolution of many new species within these groups. Some new insect groups, such as bees and butterflies, originated after flowering plants, their members developing mouthpart structures and behaviour specialized for pollination. In extreme cases, a plant is completely dependent on one insect species for pollination, and the insect is completely dependent on one plant species for food. Such tight interdependency occurs rarely but is well documented in yuccas/yucca moths, senita cacti/senita moths. In all three insects, females lay eggs in the flowers, and their young hatch later to feed on the mature fruit and its contents. Females ensure that the fruit develops by gathering pollen from another plant and transporting it to the stigma of the flower holding their eggs. Plants benefit greatly in outcrossed seed produced, at the small cost of some consumed fruit and seeds, and the insects benefit greatly from the food supply for developing larvae at the small cost of transporting pollen the short distances between plants.

Pollinating agents, whether biotic or abiotic, have exerted strong selection on all aspects of the flower, resulting in the evolution of tremendous floral diversity. This diversity has been distilled into a small number of characteristic pollination syndromes.

Pollination by beetles selects usually for white colour, a strong fruity scent, and a shallow, bowl-shaped flower. Bees select for yellow or blue/purple colorings, a landing platform with colour patterns that guide the bee to nectar (often reflecting in the ultraviolet range of the spectrum), bilateral symmetry, and a sweet scent. Butterflies select for many colors other than yellow, a corolla (petal) tube with nectar at the base, and the absence of any scent. Moths in contrast select for nocturnally opening flowers with a strong scent and drab or white colour, and also a tube with nectar at the base. Bats select also for nocturnally opening flowers, but with a strong musky scent and copious nectar, positioned well outside the foliage for easy access, and drab or white colour. Hummingbirds select for red or orange flowers with no scent, copious nectar production, and a corolla tube with nectar at the base. Other pollinating birds that do not hover while feeding select for strong perches and flowers capable of containing copious nectar (tubes, funnels, cup shapes).

Wind as a pollinating agent selects for lack of colour, scent, and nectar; small corolla; a large stigmatic surface area (usually feathery); abundantly produced, buoyant pollen; and usually erect styles and limp, hanging stamens. In addition there is great floral diversity within any of these syndromes, arising from the diverse evolutionary histories of the member plant species.

Selfing and Outcrossing

Most flowering plant species reproduce primarily by outcrossing, including the great majority of trees, shrubs, and perennial herbs. Adaptations that prevent self-fertilization include self-incompatibility (genetic recognition and blocking of self-pollen) and dioecy (separate male and female individuals). Adaptations that reduce the chances of self-pollination in hermaphrodites include separation of the anthers and stigma in space (herkogamy) or time (dichogamy). In many species, both self-incompatibility and spatiotemporal separation of the sex organs occur.

The ability to produce seeds by selfing, however, is advantageous in situations where outcrossing pollination is difficult or impossible. These include harsh environments where pollinators are rare or unpredictable, and regularly disturbed ground where survivors often end up isolated from each other. Selfing is also cheaper than outcrossing, because selfers can become pollinated without assistance from animals and therefore need not produce large, attractive flowers with abundant nectar and pollen rewards.

Most primarily selfing species are small annuals in variable or disturbed habitats, with small, drab flowers. Most desert annuals and roadside weeds, for example, are selfers. The evolutionary transition from outcrossing to near-complete selfing has occurred many times in flowering plants. Outcrossing and selfing species differ in their evolutionary potential. Outcrossers are generally more genetically diverse and produce lineages that persist over long periods of evolutionary time, during which many new species are formed. Selfers, however, are less genetically diverse and tend to accumulate harmful mutations. They typically go extinct before they have an opportunity to evolve new species.

Asexual Reproduction

The ability to produce new individuals asexually is common in plants. Under appropriate experimental conditions, nearly every cell of a flowering plant is capable of regenerating the entire plant. In nature, new plants may be regenerated from leaves, stems, or roots that receive an appropriate stimulus and become separated from the parent plant. In most cases, these new plants arise from undifferentiated parenchyma cells, which develop into buds that produce roots and shoots before or after separating from the parent. New plants can be produced from aboveground or belowground horizontal runners (stolons of strawberries, rhizomes of many grasses), tubers (potato, Jerusalem artichoke, dahlia), bulbs (onion, garlic), corms (crocus, gladiola), bulbils on the shoot (lily, many grasses), parenchyma cells in the leaves (Kalanchoe, African violet, jade plant) and inflorescence (arrowhead).

Vegetative propagation is an economically important means of replicating valuable agricultural plants, through cuttings, layering, and grafting. Vegetative reproduction is especially common in aquatic vascular plants (for example, surfgrass and eelgrass), from which fragments can break off, disperse in the current, and develop into new whole plants. A minority of flowering plants can produce seeds without the fusion of egg and sperm (known as parthenocarpy or agamospermy). This occurs when meiosis in the ovule is interrupted, and a diploid egg cell is produced, which functions as a zygote without fertilization. Familiar examples include citrus, dandelion, hawkweed, buttercup, blackberry/raspberry, and sorbus. Agamospermous species are more common at high elevations and at high latitudes, and nearly all have experienced a doubling of their chromosome number (tetraploidy) in their recent evolutionary history. These species experience evolutionary advantages and disadvantages similar to those of selfers.

Evolutionary Significance of Plant Reproduction Strategies

The attractive, colorful, and unique features of the most abundant and diverse group of land plants—the flowering plants—are believed to have evolved primarily to maximize the efficiency and speed of outcross reproduction. Each major burst of angiosperm evolution was a coevolutionary episode with associated animals, primarily insects, which were exploited to disperse pollen and seeds in ever more efficient and diverse ways.

The first major burst of flowering plant evolution was the appearance of the closed carpel together with showy flowers that were radially symmetrical. The closed carpel prevented self-fertilization through recognition and blocking of self pollen within the specialized conducting tissue of the style. Insects attracted to the showy flowers carried pollen between plants less wastefully than wind, and the radial symmetry accommodated insects of many sizes and shapes.

The second major burst was the appearance of bilaterally symmetrical flowers, which happened independently in many groups of plants at the same time that bees evolved. Bilateral symmetry forced bees to enter and exit flowers more precisely, promoting even more efficient outcross pollen transfer.

The third major burst of flowering plant evolution was the appearance of nutritious, fleshy fruits and seeds, coincident with a diversification of birds and rodents. The exploitation of vertebrates for fruit and seed dispersal resulted in less haphazard transport of offspring to neighboring populations of the same species (also visited as a food source), thereby reducing the chances that progeny inbreed with their siblings and parents and providing more assurance than wind currents that they find good habitat and unrelated mating partners of the same species.

REPRODUCTIVE SYSTEMS AND EVOLUTION IN VASCULAR PLANTS

Differences in the frequency with which offspring are produced asexually, through self-fertilization and through sexual outcrossing, are a predominant influence on the genetic structure of plant populations. Selfers and asexuals have fewer genotypes within populations than outcrossers with similar allele frequencies, and more genetic diversity in selfers and asexuals is a result of differences among populations than in sexual outcrossers. As a result of reduced levels of diversity, selfers and asexuals may be less able to respond adaptively to changing environments, and because genotypes are not mixed across family lineages, their populations may accumulate deleterious mutations more rapidly. Such differences suggest that selfing and asexual lineages may be evolutionarily short-lived and could explain why they often seem to be of recent origin.

Nonetheless, the origin and maintenance of different reproductive modes must be linked to individuallevel properties of survival and reproduction. Sexual outcrossers suffer from a cost of outcrossing that arises because they do not contribute to selfed or asexual progeny, whereas selfers and asexuals may contribute to outcrossed progeny.

Selfing and asexual reproduction also may allow reproduction when circumstances reduce opportunities for a union of gametes produced by different individuals, a phenomenon known as reproductive assurance. Both the cost of outcrossing and reproductive assurance lead to an over-representation of selfers and asexuals in newly formed progeny, and unless sexual outcrossers are more likely to survive and reproduce, they eventually will be displaced from populations in which a selfing or asexual variant arises.

The world's quarter of a million vascular plant species (Heywood and Watson, 1995) display an incredible diversity of life histories, growth forms, and physiologies, but the diversity of their reproductive systems is at least as great. In some ferns, individual haploid gametophytes produce both eggs and sperm. In others, individual gametophytes produce only one or the other. In seed plants, pollen- and ovuleproducing structures may be borne together within a single flower, borne separately in different structures on the same plant, or borne on entirely different plants. In both groups of plants, the pattern in which reproductive structures are borne influences the frequency with which gametes from unrelated individuals unite in zygotes, and it is a predominant influence on the amount and distribution of genetic diversity found in a species.

Evolutionary explanations for the diversity in mating systems once focused on differences in population-level properties associated with the different reproductive modes. Selfing or asexual plants were, for example, presumed both to be more highly adapted to immediate circumstances and to be less able to adapt to a changing environment than sexual outcrossers, and these differences were used to explain the association of different reproductive modes with particular life histories, habitats, or both (Mather, 1943; Stebbins,

1957). We now realize that to explain the origin and the maintenance of particular reproductive modes within species we must relate differences in reproductive mode to differences that are expressed among individuals within populations (Lloyd, 1965). Nonetheless, differences in rates of speciation and extinction may be related to differences in reproductive modes. As a result, understanding broad-scale phylogenetic trends in the evolution of plant reproductive systems will require us to learn more about the patterns and causes of those relationships.

MODES OF REPRODUCTION

In higher animals, meiosis produces eggs and sperm directly. The sexual life cycle of vascular plants is more complex. Multicellular haploid and diploid generations alternate. Diploid sporophytes produce haploid spores through meiosis, and those spores develop into multicellular haploid gametophytes. In pteridophytes (ferns, club mosses, and horestails) the gametophyte is free-living. In seed plants (gymnosperms and angiosperms) the female gametophyte is borne within the ovule and only the male gametophyte (pollen) leaves the structure in which it was produced. Gametophytes produce haploid egg and sperm through mitosis, and these unite to form diploid zygotes from which new sporophytes develop. Asexual reproduction in plants, as in animals, occurs when offspring are produced through modifications of the sexual life cycle that do not include meiosis and syngamy. When vascular plants reproduce asexually, they may do so either by budding, branching, or tillering (vegetative reproduction) or by producing spores or seed genetically identical to the sporophytes that produced them (agamospermy in seed plants, apogamy in pteridophytes). Vegetative reproduction is extremely common in perennial plants, especially in grasses and aquatic plants, and it can have dramatic consequences. The water-weed *Elodea canadensis*, for example, was introduced into Britain in about 1840 and spread throughout Europe by 1880 entirely by vegetative reproduction . Exclusive reliance on vegetative reproduction is, however, the exception rather than the rule. More commonly, species like white clover , reproduce both through vegetative reproduction and through sexually produced seed (Burdon, 1980).

Agamospermy is less widespread than vegetative reproduction, although it has been reported from at least 30 families of flowering plants (Gustafsson, 1947; Grant, 1981), and it is especially common in grasses and roses. Agamospermous species are often polyploids derived from hybridization between reproductively incompatible progenitors. When they have arisen many times, as in the hawk's beards (*Crepis*) of western North America (Babcock and Stebbins, 1938) or European blackberries (*Rubus*; Gustafsson, 1943), the pattern of variation makes it difficult to identify distinct lineages that can be called species.

The taxonomic distribution of apogamy in pteridophytes is not well known because of the technical difficulties associated with studying spore development. Nonetheless, Manton (1950) cites examples from at least seven genera of ferns and points out that it has been known for more than a century that apogamy can be experimentally induced in many other groups (Lang, 1898).

When vascular plants reproduce sexually, the reproductive structures may be borne in many different ways. In some pteridophytes, like the club moss *Selaginella*, and in all seed plants, eggs and sperm are produced by different gametophytes. In other pteridophytes a single gametophyte may produce both eggs and sperm, as in most ferns. Even when eggs and sperm are produced on the same gametophyte, however, zygotes most frequently are formed through union of eggs and sperm from different gametophytes (Soltis and Soltis, 1992). Differences in the time at which male and female reproductive structures form often are reinforced by antheridiogens released by gametophytes in female phase that induce nearby gametophytes to remain in male phase (Döpp, 1959). The antheridiogen system of *Cryptogramma crispa*, for example, appears to enforce outcrossed reproduction even though individual gametophytes are developmentally capable of producing both eggs and sperm (Parajón *et al.*, 1999).

Eggs and sperm are produced by different gametophytes in flowering plants, but anthers and stigmas most often are borne in a single flower. Despite the apparent opportunity for self-fertilization, zygotes most frequently are formed through the union of eggs and sperm derived from different plants (Barrett and Eckert, 1990). Genetically determined selfincompatibility mechanisms appear to have evolved several times in flowering plants (Holsinger and Steinbachs, 1997), but differences in the time at which pollen is released and stigmas are receptive within a flower and spatial separation between anthers and stigmas promote outcrossing, even in many self-compatible plants (Bertin, 1993; Chang and Rausher, 1998). In some species of flowering plants a polymorphism in stigma height is associated with a complementary polymorphism in anther height, a condition known as heterostyly. Short anthers are found in flowers with long stigmas and vice versa. Darwin (1877) described the classic example of this system in *Primula veris*. In that species as in many others, morphological differences are associated with compatibility differences that allow pollen derived from short anthers to germinate only on short stigmas and pollen derived from long anthers to germinate only on long stigmas.

In other species of flowering plants and in all gymnosperms sexual functions are separate from one another. Either pollen and ovules are produced in different structures on the same plant (monoecy) or they are produced on different plants (dioecy). The separation of sexual functions also may be associated with physiological and ecological differences between the sexes, as in *Siparuna grandiflora* in which females and males have different patterns of distribution within populations (Nicotra, 1998).

CONSEQUENCES OF REPRODUCTIVE SYSTEMS

The genetic structure of a species comprises the identity and frequency of genotypes found within populations and the distribution of genotypes across populations.

The reproductive system has long been recognized as a predominant influence on the genetic structure of plant species. Asexual progeny are genetically identical to the individuals that produced them, except for differences caused by somatic mutation. Selfed progeny may differ from their parent as a result of segregation at heterozygous loci, but selfing usually produces far fewer genotypes among offspring than outcrossing. As a result, fewer genotypes usually are found in populations in which either form of uniparental reproduction is common than in those in which outcrossing is the norm.

Among sexually reproducing species, selfers have populations with a smaller and more variable effective sizes (Schoen and Brown, 1991) and with less exchange of alleles among individuals within and among populations. As a result, selfing species are usually more homozygous than close relatives and have fewer genotypes per population than outcrossers. They also typically have fewer polymorphic loci and fewer alleles per polymorphic locus than closely related outcrossers (Brown, 1979; Gottlieb, 1981).

In addition, the diversity found within selfing species is more a result of differences among populations than of differences among individuals within populations. Over 50% of the allozyme diversity found in selfers is attributable to differences among populations, whereas only 12% is attributable to differences among populations in outcrossers (Hambrick and Godt, 1989). Allozyme and restriction site analyses of chloroplast DNA (cpDNA) in *Mimulus* (Scrophulariaceae) and nucleotide sequence analyses of introns associated with two nuclear genes in *Leavenworthia* (Brassicaceae) illustrate the dramatic impact mating systems can have on the genetic structure of plant species. In *Mimulus* both allozyme and nucleotide sequence diversity in a selfing species are only one-fourth that of a closely related outcrossing species (Fenster and Ritland, 1992). The lower nucleotide diversity in cpDNA might not be expected, because it is maternally inherited, but in highly selfing species background selection against deleterious alleles at nuclear loci can substantially reduce diversity in both nuclear and cytosolic genomes (Charlesworth *et al.*, 1995; Charlesworth *et al.*, 1997). In *Leavenworthia* populations of selfers are composed almost entirely of a single haplotype at each of the two loci studied, and each population is characterized by a different haplotype. In outcrossers, on the other hand, individuals belonging to the same population are only a little more similar to one another than were individuals belonging to different populations (Liu *et al.*, 1998, 1999). In addition, balancing selection appears to be responsible for maintaining an electrophoretic polymorphism at the locus encoding phosphoglucose isomerase in outcrossing species of *Leavenworthia* (Filatov and Charlesworth, 1999). Thus, selfers may have lower individual fitness than outcrossers, because they are genetically uniform at this locus.

The consequences of asexual reproduction are in some ways similar to those of selfing. In a strictly asexual population there is no exchange of genes among family lines, just as there is none within a completely selfing population. In contrast to selfers, however, asexual genotypes reproduce themselves exactly, except for differences caused by somatic mutation. Thus, the frequency of heterozygotes can be large in asexual populations even if the number of genotypes found is quite small, especially because many apogamous or agamospermous plants are derived from products of hybridization (Manton, 1950; Stebbins, 1950; Grant, 1981). Agamospermous *Crepis* in western North America, for example, are polyploids derived from hybridization between different pairs of seven narrowly distributed diploid progenitors (Babcock and Stebbins, 1938), and local populations are composed of relatively few genotypes. Moreover, most of the genetic diversity in the entire set of agamospermous species, which are facultatively sexual, is attributable to multiple origins rather than sexual recombination (Whitton, 1994; Holsinger *et al.*, 1999).

Although asexual populations are virtually guaranteed to have many fewer genotypes than sexual populations with similar allele frequencies, the number of genotypes within a population can still be quite large. Allozyme studies revealed between 15 and 47 clones in populations of the salt-marsh grass *Spartina patens* on the east coast of North America and 13–15 clones of the daisy *Erigeron annuus* (Hancock and Wilson, 1976). When the number of genotypes per population is large, however, most genotypes are found in only one population and only a few are found in more than two or three populations (Ellstrand and Roose, 1987). When the number of genotypes per population is small, as it often is in agricultural weeds, each one may be quite widespread.

More than 300 distinct forms of skeleton wire-weed, *Chondrilla juncea,* are found in Eurasia and the Mediterranean, but none is widespread. In Australia, however, only three forms are found, but each is widespread and the species is a serious agricultural pest (Hull and Groves, 1973; Burdon *et al.,* 1980).

In addition to effects on variation at individual loci, both selfing and asexuality may reduce the ability of populations to respond to a changing environment via natural selection, because they reduce the amount of genetic variability in populations. By exposing recessive alleles to selection, selfing may promote the loss of currently deleterious alleles that would be adaptively advantageous in other environments. In fact, selfers may maintain as little as one-fourth of the heritable variation outcrossers would maintain in a population of comparable size (Charlesworth and Charlesworth, 1995). In addition, by preventing gene exchange among family lineages, selfing and asexual populations reduce the diversity of genotypes on which natural selection can act.

In fact, the proportion of variation caused by differences among individuals within a family is expected to decline almost linearly as a function of the selfing rate in populations. In completely selfing populations, virtually all genetic differences are differences among maternal families. In outcrossing populations, genetic differences within maternal families are expected to be almost as great as those among maternal families (Holsinger and Steinbachs, 1997).

In both selfing and asexual species, therefore, the genetic structure of their populations may limit their ability to respond adaptively to natural selection. Because the causes of this constraint are the same for both types of uniparental reproduction, it is convenient to refer to it as the uniparental constraint.

Empirical analyses of phenotypic variation are consistent with the patterns of variation predicted by uniparental constraint in asexual populations (Ellstrand and Roose, 1987), but less so in populations of selfers.

Comparisons of closely related outcrossers and selfers in *Phlox,* for example, found that outcrossers had more among family variation in 11 of 20 morphological traits than selfers (Clay and Levin, 1989), although the analysis did not distinguish between genetic and environmental effects on morphological differences. A similar study in *Collinsia heterophylla* (Scrophulariaceae), however, estimated genetic components of variance and found no relationship between genetically estimated rates of selfing in populations and the partitioning of genetic variance within and among families (Charlesworth and Mayer, 1995).

In addition to reducing the ability of populations to respond to a changing environment via natural selection, both selfing and asexuality also reduce a population's effective size (Pollack, 1987). As a result, deleterious alleles that would have little chance of drifting to fixation in an outcrossing population may be effectively neutral in one that is mostly or completely self-fertilizing.

If such an allele is fixed and if it also reduces the reproductive potential of the population, the population will become smaller, making it possible for alleles that are even more deleterious to drift to fixation. This autocatalytic process, the "mutational meltdown," can lead to population extinction and can do so much more easily in completely selfing or asexual populations than in outcrossing ones. It appears, however, that a small amount of outcrossing in primarily selfing species can greatly retard the rate at which the process occurs.

Both the uniparental constraint and mutational meltdown hypotheses make an important prediction: an obligate selfing or obligate asexual lineage will be more short-lived than an otherwise comparable sexual outcrossing lineage. Unfortunately, phylogenetic analyses of the distribution of selfing and asexuality in plants are too few to allow us to assess this prediction directly. It is, however, a botanical commonplace that selfing species are often derivatives of outcrossing progenitors (see the discussion of selfers in *Arenaria* and *Linanthus* below, for example). Because derivatives must be younger than their progenitors, the average age of selfing species is probably less than that of outcrossing species, suggesting that selfers are also more short-lived.

EVOLUTION OF REPRODUCTIVE SYSTEMS

Explanations for the diversity of reproductive systems in flowering plants in terms of a tradeoff between short-term adaptive benefit and long-term flexibility were attractive, in part, because they emphasized the synergistic role of reproductive systems in plant evolution. Because of their impact on the amount and distribution of genetic variation within and among populations, reproductive systems can play an important role in determining the pattern and extent of population responses to natural selection on many other traits. As we have just seen, both obligate selfers and obligate asexuals are expected to harbor less genetic variability and to accumulate deleterious mutations more rapidly than sexual outcrossers, which could limit their ability to respond adaptively to environmental change. As a result, we would expect selfing and asexual lineages of plants to be relatively short-lived. Nonetheless, both selfers and asexuals have evolved repeatedly, and it is vital that we understand the circumstances under which they have evolved.

Failure to recognize the frequency with which self-fertilization has evolved, in particular, has led to many taxonomic mistakes. In *Arenaria* (Caryophyllaceae), for example, *A. alabamensis* is a self-pollinating derivative of *A. uniflora*, and populations of *A. alabamensis* in Georgia are independently derived from those in North and South Carolina. Because the floral features that distinguish *A. alabamensis* from *A. uniflora* are convergently derived, both sets of populations are now included within *A. uniflora* (Wyatt, 1988). *Linanthus* sect. *Leptosiphon* provides an even more striking example. Self-fertilization evolved independently at least three times in this group of 10 taxa, and one selfing taxon (*Linanthus bicolor*) appears to have three separate origins . Despite the possible impacts that self-fertilization might have on longterm persistence of populations, we have known for more than 35 years that self-fertilization "can evolve only because of a selective advantage before fertilization" (Lloyd, 1965). Moreover, there are only two types of advantage that self-fertilization can provide. It can increase reproductive success when lack of pollinators or inefficient pollen transfer limits reproductive success (reproductive assurance), or it can increase success as a pollen parent when pollen devoted to selfing is more likely to accomplish fertilization than pollen devoted to outcrossing (automatic selection) (Holsinger, 1996). Unfortunately, we are not yet able to predict when selfing will provide either reproductive assurance or an automatic selection advantage.

In *Aquilegia formosa* (Ranunculaceae), for example, pollinator exclusion does not affect seed set, suggesting that individuals are able to selfpollinate to ensure seed set. Hand pollinations increase seed set relative to open-pollinated controls, suggesting that pollen transfer limits reproductive success. Nonetheless, open-pollinated, emasculated flowers set as much seed as open-pollinated, unemasculated flowers, demonstrating that self-pollination provides little reproductive assurance in a species where both the capacity for autonomous self-pollination and pollenlimited seed set exist (Eckert and Schaefer, 1998).

The automatic selection advantage of self-fertilization, first pointed out by Fisher (1941), can arise because in a stable population an outcrossing individual will, on average, serve as ovule parent to one member of the next generation and as pollen parent to one other. A selfing individual in the same population will, however, serve as both ovule and pollen parent to its own selfed progeny and as pollen parent to one outcrossed progeny of another individual in the population.

Thus, an allele promoting self-fertilization has a 3:2 transmission advantage relative to one promoting outcrossing. So alleles promoting self-fertilization are expected to spread, unless selfed progeny suffer a compensating disadvantage in survival or reproduction.

Fisher's argument assumes that morphological changes promoting self-fertilization do not diminish the selfer's ability to serve as an outcross pollen parent. The extent to which selfing reduces an individual's contribution to the outcross pollen pool is referred to as pollen discounting (Holsinger *et al.*, 1984).

Relatively few attempts have been made to measure the extent of pollen discounting in plant populations. In one experiment in *Eichhornia paniculata* (Kohn and Barrett, 1994) and another in *Ipomoea purpurea* (Convolvulaceae) (Rausher *et al.*, 1993) selfers were actually more successful as outcross pollen parents than outcrossers.

In observations derived from natural populations of *Mimulus* (Ritland, 1991), selfers appeared not to contribute any pollen to the out- cross pollen pool. Because the selfers were morphologically quite different from the outcrossers in *Mimulus* and much less so in *E. paniculata* and *I. purpurea*, it may be that differences in the extent of pollen discounting are related to differences in floral morphology. This would be consistent with the observation that pollinator movement within multiple-flowered influorescences led to observable differences in the degree of pollen discounting in other experiments on *E. paniculata* (Harder and Barrett, 1995).

Just as the forces favoring evolution of self-fertilization, reproductive assurance, and automatic selection are well-known, so also is the primary force opposing its spread, inbreeding depression. Thomas Knight pointed out more than 200 years ago that the selfed progeny of garden peas are less vigorous and fertile than are outcrossed progeny (Knight, 1799). The impact that inbreeding depression has on the evolution of self-fertilization is, however, more complex than might be expected. The fate of a variant causing an increase in the rate of selfing depends not only on the magnitude of inbreeding depression, but also on the genetic basis of inbreeding depression, and on the magnitude of the difference in selfing rates that the variant induces.

The complexity arises because different family lineages within a population may exhibit different degrees of inbreeding depression. Because selection among family lines is an important component of natural selection in partially self-fertilizing populations, inbred families (those with a high frequency of alleles promoting self-fertilization) may show less inbreeding depression than less inbred families. If the extent of the association between family inbreeding depression and mating system is strong enough, selfing variants may spread even in the face of high population inbreeding depression (Holsinger, 1988).

The extent of associations between genetic variants affecting the mating system and levels of inbreeding depression in natural populations is not known. If the genomic rate of mutations to recessive or nearly recessive lethals is sufficiently high, levels of inbreeding depression are relatively insensitive to selfing rates (Lande *et al.*, 1994), which will cause even families that differ substantially in their selfing rate to have similar levels of inbreeding depression.

The relationship between within-population mating system differences and inbreeding depression also may be weak if mating system differences are polygenically controlled (Schultz and Willis, 1995). Experimental results are mixed. In *Lobelia siphilitica* (Campanulaceae) no differences in inbreeding depression could be found between females, which must outcross, and hermaphrodites, which self to some extent (Mutikainen and Delph, 1998), whereas in *Gilia achilleifolia* (Polemoniaceae) individuals with anthers and stigmas well separated have greater amounts of inbreeding depression than those in which anthers and stigmas are not well separated (more selfing) (Takebayashi and Delph, 2000).

Ideas about the evolution of agamospermy and apogamy in plants tend to focus on the genetic consequences of agamospermy and the mechanisms by which it might arise (Mogie, 1992). In flowering plants, for example, agamospermous reproduction resulting from asexual development of gametophytic tissue is almost invariably associated with polyploidy.

Whitton (1994) suggests that this correlation arises because the same process, formation of unreduced female gametophytes, contributes both to agamospermous reproduction and to the origin of polyploids.

Although these arguments may shed light on the evolutionary correlates of agamospermy, they shed no light on the process by which a genetic variant promoting agamospermy is able to establish itself within populations. Fortunately, it is easy to construct arguments parallel to those for the automatic selection advantage of self-fertilization to show why a similar advantage might accrue to asexual plants in a population of hermaphroditic outcrossers.

In a stable population of hermaphrodites, each outcrosser will replace itself, serving once as a seed parent and once as a pollen parent to the outcrossed progeny of another individual. Suppose a genetic variant that causes complete agamospermy is introduced into this population and that this variant has no effect on the pollen production of individuals carrying it.

Then an agamospermous individual will replace itself with agamospermous seed, but it also will serve as pollen parent to the outcrossed progeny of sexual individuals. In short, some of the seed progeny of sexuals will carry the genetic variant causing agamospermy and will be agamospermous themselves, whereas all of the seed progeny of agamosperms also will be agamospermous. Thus, agamospermy has an automatic selection advantage over outcrossing, and it will tend to spread through populations, unless agamosperms have a compensating disadvantage in survival and reproduction relative to outcrossers. I am not aware of studies that investigate the extent of the automatic selection advantage agamosperms might have in natural populations.

THE COST OF SEX

Mathematical analyses of models for the evolution and maintenance of sexual reproduction suggest that asexuals can be favored either because they avoid the "cost of males" or because they avoid the "cost of meiosis" (Williams, 1975; Maynard Smith, 1978).

The cost of males arises because the number of females in a population more often limits its rate of population growth than the number of males, a consequence of Bateman's principle (Bateman, 1948). As a result, an asexual population composed entirely of females may have a higher intrinsic rate of increase and therefore displace an otherwise equivalent sexual population with separate sexes.

In hermaphrodites, however, the cost of males will exist only when vegetative reproduction allows individuals to produce more offspring per unit of resource than reproduction through seed (Lively and Lloyd, 1990) or when selfers or agamosperms are able to divert resources from pollen to seed production (Schoen and Lloyd, 1984).

The automatic selection advantage of selfers and agamosperms often is attributed to the cost of meiosis. More careful analysis of the similarities between the evolution of selfing and the evolution of agamospermy suggests that cost of meiosis is not an apt description for the forces governing either process.

The phrase cost of meiosis refers to the idea that the genetic coefficient of relatedness between individuals and their outcrossed offspring is smaller than the coefficient of relatedness would be between those same individuals and their selfed or asexual offspring. Notice, however, that in a population of complete selfers an agamosperm would not have an automatic selection advantage over selfers because the pollen it produces would not fertilize any ovules. The asexual progeny of an agamosperm are genetically identical to their parent (barring rare somatic mutation) and the selfed progeny of a selfer are genetically variable to the extent that there is segregation at heterozygous loci. Thus, the asexual progeny of an agamosperm are more closely related to their parent than the sexual progeny of a selfer. Nonetheless, the relative fitness of selfed and agamospermous offspring will determine the outcome of natural selection, not the extent to which selfed or agamospermous progeny resemble their parents.

These observations suggest that cost of outcrossing is a better phrase to describe the automatic selection advantage of selfers and agamosperms relative to sexual outcrossers. The cost of outcrossing arises because selfers and agamosperms can serve as pollen parents of progeny produced by sexual outcrossers, but outcrossers are prevented from serving as pollen parents to the selfed progeny of selfers and the asexual progeny of agamosperms. So long as pollen devoted to selfing is more likely to accomplish fertilization than pollen devoted to outcrossing (Holsinger, 1996) and so long as agamosperms are able to serve as pollen parents to the outcrossed progeny of other individuals, selfing and agamospermy will be overrepresented in newly formed progeny of the next generation. Unless, natural selection against selfed or agamospermous progeny is sufficiently strong, the cost of outcrossing will cause the frequency of outcrossers to decline.

PLANT REPRODUCTION SYSTEM

The plants that sexually reproduce have the reproductive structures called the flowers. The flower is a condensed shoot with the nodes present very close to each other. The different parts of the plant are attached to the nodes. All the structures present at one node are collectively called the whorl. The first or the outermost two whorls are called the non-reproductive whorls. They are the calyx and corolla. The inner two whorls are androecium and gynoecium, the reproductive whorls.

Parts of a Flower

Calyx: is the outermost and most often green in colour. The individual units of calyx are called the sepals. It protects the inner whorls at bud stage.

Corolla: is the next inner whorl and is often coloured brightly. The individual units of corolla are called petals. They serve to attract bees, birds, etc which are the agents of pollination.

Androecium is the male reproductive part of the flower. The individual units of androecium are called the stamens. Each stamen has a thread-like filament at the free end of which is attached the four-lobed anther.

Structure of Anther

The anther has four pollen-sacs, one in each lobe. The pollen-sacs contain cells called the microspore mother cells (MMC) or the pollen mother cells. MMCs undergo mieosis to produce four haploid microspores, also called the pollen grains. Each microspore divides once mitotically to produce two male gametes or sperm cells. Thus, each microspore mother cell produces 8 sperm cells.

Gynoecium is the female reproductive part of the flower. The individual units are called the carpels or pistils.

A flower may have one to many carpels, either fused or free. Each carpel is made up of the basal ovary, middle style and the upper stigma. The ovary is the chamber where there are many ovules that are attached to an axis. Each ovule consists of a haploid egg and other associated cells. The stigma is a sticky structure that receives the pollen grains. The style is hollow and provides a passage for the male gametes to reach the female gametes, the eggs.

Pollination

Transfer of pollen grains to the stigma is called pollination. If the pollen grains are transferred to the stigma of the same flower, the pollination is called self-pollination or autogamy. If the pollen grains are transferred to the stigma of another flower of the same species, the pollination is called cross-pollination or allogamy.

The anthers on maturity burst open with force and this is called dehiscence. This releases the pollen grains with force which are then carried by wind and water to other plants. In other plants, the flowers are brightly coloured and scented to attract the birds, bees, etc. The insect or the bird enters the flower to suck the nectar produced by glands at the base of the flower. The pollen grains present on the dehisced or open anther, stick to the legs or abdomen. When the same insect visits other flowers the pollen grains are transferred to the stigma of those flowers.

Fertilisation

The tip of the tube contains the male nuclei. The tube grows and enters the ovule where it bursts at the tip releasing the male gametes. One of the male gametes fuses with the egg, the female gamete. The fusion of the male gamete with the female gamete is called fertilisation. This results in the formation of zygote that is diploid. The zygote develops into the embryo. The other male gamete fuses with the polar nuclei. This results in the formation of a triploid nucleus called the endosperm nucleus. Since the process of fertilisation involves two fusions, it is called double fertilisation.

The divisions of the endosperm nucleus result in the formation of the endosperm that nourishes the growing embryo. The ovule then becomes the seed and the ovary changes into fruit.

Plant Reproduction

The way plants reproduce is a totally fascinating process. Out of the more than 300,000 different kinds of plants, more than half are seed plants. Seed plants make their own seeds from which new plants grow. Other ways plants can make new plants are from spores, rhizomes, bulbs, tubers, corms, cuttings, grafts, and buds. Seeds are made by flowers in some plants and by cones in other plants.

Flower parts contain specialized cells, called ovules, which have the job of making the seeds. Different kinds of flowers have many different parts.

Basically, most plants—trees, vegetables, and even grass—sprout flowers. Some of these flowers are beautiful and colorful while others are so tiny one can hardly see them, but all plants produce new plants.

Botanists classify the reproduction of plants as ones with perfect and imperfect flowers. Perfect flowers are the ones that contain both the male and female parts. Imperfect flowers are the ones that contain only male or only female parts. So, of course, these different flowers have to find different ways to propagate themselves.

The stamen is the male part of a flower and has an anther on a stalk. The anther's job is to produce the pollen. Then there is the pistil, or female part. The pistil contains the flower's ovary, the style, and the stigma. Inside the ovary are the ovules. Each ovule contains an egg cell. So when an egg cell joins with a pollen cell, a seed may be produced.

The ovary is at the base of the plant. The stigma is at the top, and between them is the style. So, the pollen from the anther has to land on the stigma. When that happens, the pollen germinates and makes its way down into the ovary. Inside the ovary are the ovules, which contain egg cells. When an egg cell joins with a pollen cell, a seed may be produced. Then the ovary becomes a fruit. So a fruit is a mature ovary holding the seeds. In fruits, like the pears, the pear's core is the seedpod of the pear tree. Acorns are the seedpods of oak trees. And the tomato is the seedpod of the tomato plant. A flower garden is alive with insects. They are all after the sweet liquid called nectar, which the flowers produce, near the bottom of the pistil. Flowers produce the nectar to draw the insects to them. Insects—like butterflies, wasps, and bees—come to get the nectar. Some, like butterflies and wasps, use the nectar for food; some, like bees, make it into honey and store it in their hives.

If you look closely at the body of a butterfly or bee, you will notice it is covered with tiny hairs. As they land on the flower to collect the nectar, some of the pollen is brushed on these hairs. Then the insect carries the pollen to the next flower, and pollination takes place.

Some plants, like corn, or flowers of an oak tree, rely on the wind to spread their pollen. So these kinds of plants don't produce pretty flowers since they don't need to attract insects to come and pollinate them.

Some plants need long tongues to pollinate them. Their flowers are usually tubular, like the trumpet creeper vine. These kinds of flowers are best pollinated by hummingbirds or by moths that have long tongues. And flowers that open at night do so because they need the specialized pollination provided by certain moths. Even the colour of plants makes a difference. Birds, like the hummingbird, usually pollinate plants that are red. Night insects pollinate white and pale yellow flowers, which are seen better at night.

Trees that have needle-like leaves are called conifers. Conifers produce cones instead of flowers. Under each scale of a cone is a part that produces either pollen or an ovule. But a cone cannot produce both. The very small cones produce the pollen in the spring.

The wind carries the pollen to the larger cone, which has the ovule producing part. Then, the pollen producing cones fall off while the fertilized ovules grow larger. Eventually, they open and release the seeds, and some of the seeds sprout. Some plants, like tulips, grow from bulbs. Bulbs produce plants very easily by reproducing underground.

Plants can be grown without seeds by cuttings. A cutting can be from any part of the fully-grown plant, and when it is placed in soil, it may produce a new plant. You can increase your own houseplants by this method, and here is the way to do it. With a sharp knife, cut a piece of stem from the houseplant you want to use.

Pull off any leaves or flowers from the stem's lower part. Stick the stem in some clean, moist sand. Cover the flowerpot with a clear, plastic bag. Place in a well-lit place. After a few weeks, lift the stem out of the sand. Are the roots beginning to grow? Now you can plant it in a pot with potting soil. Keep the soil moist, and soon, the new plant will begin to grow. When cuttings from trees and other woody stems are attached to another plant, this is known as grafting. Grafts are another way of growing plants without using seeds.

There are a variety of ways plants produce new plants. To find out more about this fascinating process, you may want to read books on the subject or do some further research on the Internet.

GENETIC VARIATIONS IN PLANTS

Variation

If you have grown Fast Plants® you will have noticed variation among the individual plants within the group. Variation can range from a little to a lot. This article is designed to help teachers and students understand the basis of the variation they observe in their plantings.

Variation is one of the fundamental characteristics of life. All organisms exhibit some variation among individuals. Understanding the ways that variation is manifested in organisms, how it comes to be expressed through the development of the individual, and how it is transmitted from one individual to the next generation of individuals are central themes in biology.

Working with Fast Plants® will enrich a student's understanding of variation. By observing the growth and development of a Fast Plants® through the various stages in the life cycle, students will become aware of many visible features, or phenotypes, that make up the organism. Only, however, upon close observation of a population of plants, will they become aware that the characteristics observed on one plant vary more or less on other plants. Such is the nature of variation.

Phenotypic Variation

Phenotypic variation, e.g. plant height at a particular stage of development, is considered to be the expression of the genetic makeup (genotype) of the individual as it interacts with the environment. Variation in plant height among individuals in a population is therefore due to variation in the interaction between the genotype and the environment, as expressed through the development of each individual plant.

DESCRIBING AND OBSERVING PHENOTYPIC VARIATION

In order to be useful in an experiment the phenotype must be described using terms that are widely understood and easily communicated. For these reasons scientists have agreed upon various standards or descriptors to describe characteristics in the natural world. Descriptors take many forms, see WFPID Observing and Describing. The choice of how to describe what you observe is important, because it will determine the kinds of descriptors used and establish the basis for recording, analyzing and communicating results.

Environmental Variation

Much can be learned about the role of light, temperature, and nutrients on plant development from experiments in which one or more environmental parameters are changed. Because Fast Plants® are highly responsive to changes in the environment, they are ideal for examining the role of the environment on the expression of phenotypic variation. Although Fast Plants® are able to grow within a wide range of environmental conditions, for most investigations it is recommended that they be grown under uniform and ideal conditions. In this way variation arising from sub-optimal conditions of environment will be minimized. The Wisconsin Fast Plants® Information Document (WFPID) Understanding the Environment describes how to provide and maintain the various physical, chemical and biotic components of the environment that are most suitable for Fast Plants®.

Genotypic Variation

How can students use Fast Plants® to investigate the contribution of the genotype to the phenotype? Fast Plants®, rapid cycling Brassica rapa, are genotypically variable in that they have a genetically controlled mating system that prevents self-fertilization and favours out-crossing among individuals.

As a consequence, even seed stocks selected for uniformity of specific phenotypes and genotypes exhibit considerable variation for other traits.

The Wisconsin Fast Plants® Program has developed a number of genetic stocks of rapid cycling Brassica rapa for genetic investigations on the nature and inheritance of variation. Some WFP stocks contain distinctive mutant phenotypes, e.g., anthocyaninless plant, anl, yellow green plant, ygr1, rosette, ros, and male sterile, mst2, that are suitable for Mendelian genetics. Other stocks exhibit phenotypes whose expression may vary continually and which may be quantified as discrete or countable units, e.g., number of hairs, or as estimates of size, e.g. petite dwarf, dwf1, or of intensity of colour saturation, e.g., purple anthocyanin. These quantitative phenotypes may be conditioned by a few or many genes and normally require numerical descriptions in which the statistical notations of population size, (n), range (r), arithmetic mean (x), and standard deviation (s) are applied. Yet other stocks have been developed to combine both simply inherited mutant genotypes and quantitatively expressed phenotypes. An important part of WFP is the continuing development and improvement of seed stocks for uses in genetics.

GENETIC DIVERSITY

Genetic diversity is a level of biodiversity that refers to the total number of genetic characteristics in the genetic makeup of a species. It is distinguished from genetic variability, which describes the tendency of genetic characteristics to vary.

The academic field of population genetics includes several hypotheses and theories regarding genetic diversity. The neutral theory of evolution proposes that diversity is the result of the accumulation of neutral substitutions. Diversifying selection is the hypothesis that two subpopulations of a species live in different environments that select for different alleles at a particular locus. This may occur, for instance, if a species has a large range relative to the mobility of individuals within it. Frequency-dependent selection is the hypothesis that as alleles become more common, they become less fit. This is often invoked in host-pathogen interactions, where a high frequency of a defensive allele among the host means that it is more likely that a pathogen will spread if it is able to overcome that allele.

Importance of Genetic Diversity

There are many different ways to measure genetic diversity. The modern causes for the loss of animal genetic diversity have also been studied and identified. A 2007 study conducted by the National Science Foundation found that genetic diversity and biodiversity are dependent upon each other — that diversity within a species is necessary to maintain diversity among species, and vice versa. According to the lead researcher in the study, Dr. Richard Lankau, "If any one type is removed from the system, the cycle can break down, and the community becomes dominated by a single species."

Survival and Adaptation

Genetic diversity plays a very important role in survival and adaptability of a species because when a species's environment changes, slight gene variations are necessary to produce changes in the organisms' anatomy that enables it to adapt and survive. A species that has a large degree of genetic diversity among its population will have more variations from which to choose the most fit alleles. Increase in genetic diversity is also essential for an organism to evolve. Species that have very little genetic variation are at a great risk. With very little gene variation within the species, healthy reproduction becomes increasingly difficult, and offspring often deal with similar problems to those of inbreeding. The vulnerability of a population to certain types of diseases can also increase with reduction in genetic diversity.

Agricultural Relevance

When humans initially started farming, they used selective breeding to pass on desirable traits of the crops while omitting the undesirable ones. Selective breeding leads to monocultures: entire farms of nearly genetically identical plants. Little to no genetic diversity makes crops extremely susceptible to widespread disease. Bacteria morph and change constantly. When a disease causing bacterium changes to attack a specific genetic variation, it can easily wipe out vast quantities of the species. If the genetic variation that the bacterium is best at attacking happens to be that which humans have selectively bred to use for harvest, the entire crop will be wiped out.

A very similar occurrence is the cause of the infamous Potato Famine in Ireland. Since new potato plants do not come as a result of reproduction but rather from pieces of the parent plant, no genetic diversity is developed, and the entire crop is essentially a clone of one potato, it is especially susceptible to an epidemic. In the 1840s, much of Ireland's population depended on potatoes for food. They planted namely the "lumper" variety of potato, which was susceptible to a rot-causing plasmodiophorid called *Phytophthora infestans*. This plasmodiophorid destroyed the vast majority of the potato crop, and left tens of thousands of people to starve to death.

Coping with Poor Genetic Diversity

The natural world has several ways of preserving or increasing genetic diversity. Among oceanic plankton, viruses aid in the genetic shifting process.

Ocean viruses, which infect the plankton, carry genes of other organisms in addition their own. When a virus containing the genes of one cell infects another, the genetic makeup of the latter changes. This constant shift of genetic make-up helps to maintain a healthy population of plankton despite complex and unpredictable environmental changes.

Cheetahs are a threatened species. Extremely low genetic diversity and resulting poor sperm quality has made breeding and survivorship difficult for cheetahs -- only about 5% of cheetahs survive to adulthood. About 10,000 years ago, all but the jubatus species of cheetahs died out. The species encountered a population bottleneck and close family relatives were forced to mate with each other, or inbreed. However, it has been recently discovered that female cheetahs can mate with more than one male per litter of cubs. They undergo induced ovulation, which means that a new egg is produced every time a female mates. By mating with multiple males, the mother increases the genetic diversity within a single litter of cubs.

DIAGRAM OF A DICOT PLANT

One of the simplest methods of reproduction is when one cell divides and becomes two equal halves that will grow large enough to split again. With plants and animals that reproduce this way, each generation is identical to the one before.

Cell Dividing Diagram With One Parent Cell

Plant reproduction may be sexual, in which two parents produce a genetically different individual; or asexual, involving the propagation of plants that are genetically identical to the parent.

The Sexual Plant Life Cycle

Sexual reproduction is important in providing genetic variability. All plants that reproduce sexually must go through meiosis. Meiosis is a unique kind of cell division during which the paired sets of chromosomes present in sexually mature plants, called sporophytes, are halved. Because they have the two sets of chromosomes, one from the male parent and the other from the female, the cells of sporophytes are called *diploid.*

During this process, pairs of homologous, or identical, chromosomes, one from each parent, line up together. Crossing-over, or the exchange of genetic material between these homologous chromosomes, may occur at this time. Crossing-over is critical for producing some of the genetic variability in resulting offspring. Meiosis typically produces four haploid cells, each with one set of chromosomes, from a single diploid cell. These haploid cells become gametes-eggs and sperm. Without meiosis, sexually reproducing organisms would not have a mechanism for reducing the total chromosome number by half so that genetic variability can be introduced via crossing-over. Meiosis is a critical step that must occur prior to the fusion of sperm and egg, which restores the diploid chromosome number.

Alternating Generations

Plants undergo a two-phase cycle of sexual reproduction known as the *alternation of generation*. This sexual life cycle involves alternation of the diploid sporophyte generation with the haploid gametophyte generation. The sporophyte stage begins with fusion of an egg and sperm, which produces a zygote. The diploid zygote develops into a sporophytic plant. This sexually mature plant eventually produces meiocytes, cells that will undergo meiosis, typically resulting in four haploid cells. These haploid cells, in turn, develop into gametophytes, which eventually form sex organs where eggs and sperm are produced. The sporophyte and gametophyte stages usually are easy to tell apart; one is typically parasitic or dependent upon the other. Which generation is dominant differs in different plants. Generally, the gametophytes of primitive land plants such as mosses are dominant and larger than the sporophytes. In higher, or more advanced plants, the reverse is true. In other words, the alternation of generations is the alternation between meiocytes undergoing meiosis and the fusion of gametes that produces a zygote.

The Gametophyte Stage

The gametophyte generation normally begins with a spore (sexual spore or meiospore) and ends with a gamete. The cells of the gametophyte are usually haploid.

The Sporophyte Stage

The sporophyte generation usually begins with a zygote and ends with a meiocyte. The cells of the sporophyte are usually diploid.

Reproduction in Fern

Sporophyte development is very rapid in common garden and woodland ferns. The zygote develops an embryonic root, leaf, stem, and foot. The foot apparently aids in absorbing nutrients from the gametophyte. The stem develops slowly, but ultimately produces leaves. The first few leaves of juvenile ferns differ from those of the mature sporophyte. As growth continues, the leaves finally begin to resemble those of the adult species.

Life Cycle of a Fern Diagram.

When the minute sporophyte has become established, the gametophyte dies. The gametophyte generation in ferns, as in other seedless vascular plants, is relatively short-lived and simple. The sporophytic phase is dominant in the life cycle. As in all vascular plants, what we recognize as "the plant" is the sporophytic generation, and at maturity it develops sporangia, the spore-producing organs.

In the common ferns, sporangia are borne on the undersides of fronds, or on modified fronds. The first fronds produced in spring are sterile-that is, they lack reproductive structures. The fertile fronds arise later in the season. In some species, such as lady fern (*Athyrium filix-femina*) the sterile and fertile fronds are similar. In other species-sensitive fern (*Onoclea sensibilis*), for example-the sterile and fertile fronds are very different. The fertile regions on fronds that bear sporangia are known as *receptacles,* and the group of sporangia on a single receptacle is called a sorus. In many species, but by no means in all, the sorus is covered during development by a flap of tissue called an *indusium.* The shape and arrangement of the sori are different in different species.

Spores arise by meiosis from sporocytes. For the most part, immature sori are a pale, whitish colour, although there are exceptions, such as autumn fern (*Dryopteris erythrosora*), whose sori, which are produced in the fall, are bright red. As the spores inside each sporangium mature, they typically get darker, until they are a deep brown or black. Not all spores mature to a deep brown or black, however. Polypodiums are ripe when they are buttercup yellow; osmundas, when they are green. The sporangia of ferns is typically a thin-walled case, usually on a stalk, and surrounded by a ring of thick-walled cells known as the *annulus,* which aids in opening the sporangium when the spores are mature.

The indusium, if present, ultimately shrivels when the spores have matured, exposing the sporangia. The annulus breaks near the base of one side, tearing the sporangium apart, and arches backward. Then the annulus snaps forward abruptly, catapulting the spores into the air. Ferns produce tremendous numbers of spores, but the special requirements of most species for moisture and shade effectively reduce the number of gametophytes that ultimately develop. Spores that are deposited by air currents on sufficiently moist soil and rocks germinate within five to six days, and the gametophyte begins to develop. Most fern spores also require light for germination; those of the bracken fern (*Pteridium aquilinum*) are an exception. Early in germination, colourless rhizoidal cells form at the base of the filament of green cells. These rhizoids, or root-like hairs, absorb and carry water and nutrients to the developing gametophyte. Eventually, the gametophyte becomes heart-shaped, and in some species may be as big as half an inch in diameter. The fern gametophyte was long ago called the *prothallus* or *prothallium,* because it was known to be the precursor of the fern plant, even before its sexual function was clearly understood.

As well-nourished, bisexual gametophytes mature they develop antheridia and archegonia, normally on their undersides. The female, egg-producing organs, or archegonia, occur near the notch of the heart-shaped prothallus. The male, sperm producing organs, or antheridia, occur on the "wings" and opposite the notch. Each archegonium has a chimney-like protuberance that is flared at the top to receive the sperm. Each antheridium is a minute, capsule-like sac where the sperm grow. The cells inside the neck of a mature archegonium dissolve, creating a moist passageway through which the sperm can swim to the mature eggs. The eggs of several archegonia may be fertilized, but usually only one of the zygotes develops into a juvenile sporophyte. It is interesting to note that while bisexual prothalli may be capable of fertilizing themselves, experiments have shown that the resulting sporophytes often fail to thrive. This suggests that cross-fertilization, the union of sperm and egg from different prothalli, necessary for increasing genetic variability within species, is promoted in at least some ferns.

Asexual Reproduction

Asexual propagation can occur in various ways. Many plants produce genetically identical copies of themselves, through a mechanism referred to as asexual reproduction. However, botanists more properly refer to this mechanism as "sexual propagation" or "vegetative propagation" because many hold that only sexual reproduction should be referred to as true reproduction, since this is the only kind of propagation that results in the production of new, genetically unique individuals.

Another type of asexual propagation occurs when plants develop underground stems (or rhizomes) which grow outward, or new shoots which grow upward to form new shoots that are genetically identical to the parent. One such example is the trembling aspen (*Populus tremuloides*), which sometimes develops entire stands of trees growing out of the ground as seemingly individual stems, but are actually genetically identical and interconnected below-ground. Another example is the strawberry (*Fragaria virginiana*), although this species has its vegetative runners above-ground.

Other plants develop bulbils on their stems, which can detach, fall to the ground, and sprout to develop new plants that are genetically identical to the original one. One familiar species that does this is the tiger lily (*Lilium tigrinum*. Other plants can propagate from twigs or stem pieces that fall from the parent, then lodge into a suitable site and develop into a new plant. The crack willow (*Salix fragilis*) can spread itself along watercourses in this manner (as well as by disseminating seeds). Other non-sexual means of propagation include the production of underground bulbs, corms, and tubers that split into parts, each of which is capable of developing into a new plant. Plants that reproduce in this manner include irises and daffodils. All plant organs have been used for asexual reproduction, but stems are the most common.

Stems

In some species, stems arch over and take root at their tips, forming new plants. The horizontal above-ground stems (called stolons) of the strawberry (shown here) produce new daughter plants at alternate nodes.

Underground Stems

- Rhizomes
- Bulbs
- Corms and
- Tubers

Irises and day lilies, for example, spread rapidly by the growth of their rhizomes.

Leaves

This photo shows the leaves of the common ornamental plant Bryophyllum (also called Kalanchoë). Mitosis at meristems along the leaf margins produce tiny plantlets that fall off and can take up an independent existence.

Roots

Some plants use their roots for asexual reproduction. The dandelion is a common example. Trees, such as the poplar or aspen, send up new stems from their roots. In time, an entire grove of trees may form—all part of a clone of the original tree.

Plant Propagation

Commercially-important plants are often deliberately propagated by asexual means in order to keep particularly desirable traits (*e.g.*, flower colour, flavour, resistance to disease). Cuttings may be taken from the parent and rooted. Grafting is widely used to propagate a desired variety of shrub or tree.

All apple varieties, for example, are propagated this way. Apple seeds are planted only for the root and stem system that grows from them. After a year's growth, most of the stem is removed and a twig (scion) taken from a mature plant of the desired variety is inserted in a notch in the cut stump (the stock). So long the cambiums of scion and stock are united and precautions are taken to prevent infection and drying out, the scion will grow. It will get all its water and minerals from the root system of the stock. However, the fruit that it will eventually produce with be identical (assuming that it is raised under similar environmental conditions) to the fruit of the tree from which the scion was taken.

Apomixis

Citrus trees and many other species of angiosperms use their seeds as a method of asexual reproduction; a process called apomixis.

- In one form, the egg is formed with 2n chromosomes and develops without ever being fertilized.
- In another version, the cells of the ovule (2n) develop into an embryo instead of—or in addition to—the fertilized egg.

Hybridization between different species often yields infertile offspring. But in plants, this does not necessarily doom the offspring.

Many such hybrids use apomixis to propagate themselves. The many races of Kentucky bluegrass growing in lawns across North America and the many races of blackberries are two examples of sterile hybrids that propagate successfully by apomixis.

- The female cones of their own species (rare) or
- Those of a much more common species of cypress.

Breeding apomictic crop plants

Many valuable crop plants (*e.g.*, corn) cannot be propagated by asexual methods like grafting. Agricultural scientists would dearly love to convert these plants to apomixis: making embryos that are genetic clones of themselves rather than the product of sexual reproduction with its inevitable gene reshuffling. After 20 years of work, an apomictic corn (maize) has been produced, but it does not yet produce enough viable kernels to be useful commercially.

IMPORTANCE OF MODE OF REPRODUCTION

The mode of reproduction of a crop determines its genetic composition, which, in turn, is the deciding factor to develop suitable breeding and selection methods (what is meant is for instance: a self-ferilizing crop like wheat is bred by basically different methods than a cross-fertilizing plant like rye grass; this is not a discussion on e.g. biotechnology vs. organic work, this is all within classical, traditional breeding). Knowledge of mode of reproduction is also essential for its artificial manipulation to breed improved types. Only those breeding and selection methods are suitable for a crop which does not interfere with its natural state or ensure the maintenance of such a state. It is due to such reasons that imposition of self-fertilization on cross-pollinating crops leads to drastic reduction in their performance. Likewise, it is practically impossible to maintain permanent heterozygosity in self-fertilizing crops rendering the development of hybrids an unattractive breeding approach. Asexual propagation is another type of reproduction where any plant or part of it can be used for multiplication without even a slight genetic change from generation to generation. The methods of breeding and multiplication for such crop are thus entirely different from those of sexually reproducing crops.

For teaching purpose, plant breeding is presented as four categories: Line breeding (autogamous crops), population breeding (allogamous crops), hybrid breeding (mostly allogamous crops, some autogamous crops), clone breeding (vegetatively propagated crops). Within these categories, there are several breeding methods possible. Moreover, there is a wide spectrum of techniques and technologies to make several steps in these methods easier, faster, more precise, or allow them at all, microspore culture, embryo rescue, mutagenesis, Agrobacterium-mediated gene transfer, fingerprinting, tilling, QTL analysis etc.

SELF FERTILIZING CROPS (AUTOGAMOUS CROPS)

Certain restrictions caused the mechanisms for self fertilization (partial and full self fertilization) to develop in a number of plant species. Some of the reasons why a self fertilizing method of reproduction is so effective are the efficacy of reproduction, as well as decreasing genetic variation and thus the fixation of highly adapted genotypes. Most of the loci get fixed at a high rate; this can be ascribed to the fact that with each generation of self fertilization the rate of heterozygotes decreases by 50%. Homozygosity will thus be obtained in 5-8 generations. The 3rd reason for the efficacy of self fertilization is that in mixed stands of self and cross pollinating crops, the self fertilizing plants can donate pollen to both plant types, where the cross fertilizing plants are restricted concerning the contribution it can make to the population with regard to pollen donation.

Almost no inbreeding depression occurs in self fertilizing plants because the mode of reproduction allows natural selection to take place in wild populations of such plants. Thus, the genetically non-superior or unstable plants are removed from the population at an early evolutionary stage. Populations derived from self pollination are sometimes not as evolutionary adaptable as with other reproductive methods, but are known to utilize specific ecological niches more effectively.

Critical steps in the improvement of self fertilizing crops are the choice of parents and the identification of the best plants in segregating generations. The breeder should also have definite goals with the choice of parents. Self fertilizing cultivars are easier to maintain, but this could lead to misuse of seed. Some of the agronomically important, self fertilizing crops include wheat, rice, barley, dry beans, soy beans, peanuts, tomatoes, etc.

New new genetic variation is released in a given germplasm pool by crossing within this pool or with foreign genotypes, followed by segregation;

moreover, mutagenesis and introduction for alien DNA (transgenes) are further options to create new variation. Several selection methods are applied in self fertilizing crops, such as mass selection, single plant-, pedigree-, bulk population-, back cross-, recurrent-, as well as single seed descend (SSD) selection and the use of DH (doubled haploid) lines. In most breeding programs a combination of these methods are applied. The different selection methods can be summarized as follows:

Mass Selection

This method of selection depends mainly on selection of plants according to their phenotype and performance. The seed from selected plants are bulked for the next generation. This method is used to improve the overall population by positive or negative mass selection. Mass selection is only applied to a limited degree in self fertilizing plants and is an effective method for the improvement of land races. This method of selection will only be effective for highly heritable traits. one shortage of mass selection are the large influence that the environment has on the development, phenotype and performance of single plants. It is often unclear whether the phenotypically superior plants are also genotypically superior and strong environmental differences may lead to low selection efficacy (heritability).

Single Plant Selection (Pure Line Selection)

A variety developed by this method will be more uniform than those developed by mass selection because all of the plants in such a variety will have the same genotype. The seed from selected plants are not added together but are kept apart and used to perform offspring tests. This is done to study the breeding behaviour of the selected plants. The high uniformity in stand and performance has been stressed in the past, but the risk of highly specialized pathogens evolving is very high. More genetic variability could buffer the crop against such pathogens as well as stability of production under varied environmental conditions.

SELECTION METHODS FOR THE DEVELOPMENT OF PURE BREEDING CULTIVARS FROM CROSSES

Crosses between varieties, germplasm introduction and breeding lines are made to create new gene combinations. In the generations to follow, superior genotypes (presumably having superior genes and gene combinations) are selected and fixed in the homozygous state by means of self fertilization and selection. These selections are tested extensively with the goal of releasing the few best of them for cultivation.

Pedigree Selection

Parental lines are crossed and selection of plants with new gene combinations already takes place in the F2 generation (the generation of plants formed from (natural) selfing of the F1 hybrids). The offspring of selected populations in the generations to follow are repeatedly subjected to selection and always single plants are taken as source of new offpspring lines, until genetic uniformity is reached. Only then, seed yield of several plants per line is combined and mixed to have more seed for testing. Records are kept of the origin of the selected individuals or lines. The amount of generations of single plant and line selections, as well as selection intensities, can be varied in practice according to the crop and availability of facilities. It is usually traits with high heritability that are quick and easy to measure that are concentrated on as long as single-plant-selection is carried out (in the early phase. In the later phase, when plots are the units of assessment, traits like yield with lower heritability are assessed and taken as basis for selection. One of the main objections against this method is that the genetic variation, available for selection of quantitative traits, are drastically decreased in later generations due to the single-plant-selection carried out in early generations. Seed purification and multiplication is usually incorporated in one of the final generations of pedigree selection. This method is very labour intensive.

Breeding efficiency is one of the goals with early generation testing. This is done by early identification of superior heterogeneous populations. The early elimination of inferior populations and subsequent concentration of selection efforts within superior populations is assumed to result in increased efficiency. Accurate evaluation of heterogeneous populations is essential to the success of this method and assumes that transgressive segregants from inferior populations will not exceed selections from superior populations in performance.

Bulk Population Selection

With this method of selection the offspring from a crossing are planted at planting densities equal to commercial planting densities. During this period, which may include a number of generations, the level of homozygosity in the bulk population increases.

This method is simple and cheap and involves less work than pedigree selection in the earlier generations. It is necessary to plant large populations to ensure that the best segregates are selected when selection starts. Segregating generations are subjected to another single plant selection step. Fewer records are kept during earlier generations than with pedigree selection. This type of selection is especially carried out with crops which are usually planted at high planting densities, e.g. small grain crops.

Single Seed Selection (SD WEED)

This method was introduced as a means to transform the entire F2-generation as fast as possible into a generation of homozygous plants, albeit heterogeneous population (thus prevent the loss of variation and have the plants homozygous, i.e., tru-to-type breeding).

This method is hence used to decrease the time that is passing with genotypes not yet being homozygous. The decrease is enabled by the fact that with only one seed desired per plant, plants can be pushed to have two or three, sometimes even more seasons per calendar year. This can cause the process from the starting a breeding cycle until release of cultivar to decrease by 1-3 years. This method does not eliminate weak plants early such as in other methods, there is no provision for selection of superior plants in the F2 generation. Modification of this method is possible and record keeping is not necessary in early generations.

Doubled Haploid Method

Haploid plants may be produced by chromosome elimination in wide crosses, ovule culture or by anther and micro spore culture. Anther and micro spore culture, however, are mostly used because of its ability to produce haploid plants with much larger quantity compare with the other two methods. Stresses are usually necessary to alter the development pathways of micro spores from producing pollen to forming haploid plants. Presently, the so-called inducer method is very much used in maize breeding (hybrid maize)

The chromosome numbers of the haploid plants are doubled with the use of cochineal. Spontaneously doubled haploid plants, however, can also be produced directly from the three methods. Embryo rescue methods can be used to ensure that seed from these wide crosses or stigma culture plants survive and doesn't get aborted. This method has the potential of shortening genetic improvement cycles in comparison to pedigree or bulk methods. Like the single seed selection method, early generations are not subject to selection, but most of the lines are eliminated during the field evaluation trials.

The purpose is the same as for SSD: transform the genetic material from F1 as fast as possible into a bulk of homozytous plants. This method is very labour intensive and the most expensive of the procedures that increases the amount of generations per year. For this method to be successful, the plants must be genetically stable.

FACTORS AFFECTING PLANT REPRODUCTION

Climate, soil moisture, soil nutrients, disturbances (such as defoliation, fire, cultivation, and disease) are some of the more common things that affect the reproductive ability of plants. Tolerance versus Requirements: not all plants respond the same to stress. For example, both Agrostis tenuis and Festuca ovina have the same tolerance limits for soil pH, but in a stand of Agrostis and Festuca, Agrostis had a maximum abundance at pH 4-6 and Festuca had a maximum abundance at pH 3-4 and 6.5-7.5. There is also a difference within species. For example, 2 ecotypes of Dactylis glomerata (Norwegian and Portugese) have different temperature requirements for reproduction. In Portugal, it is a winter-grown crop (lower temps) and in Norway it is a summer crop (higher temps) and they respond differently when grown in each other's typical environment.

Plant Community Development

Plant Community - "a group of plants living under relatively similar conditions in a definite area"

Development-"to expand or bring out the potentialities; to advance from a lower to higher state" If these two are combined, then plant community development could be considered "to expand or bring out the potentialities a group of plants living under relatively similar conditions in a definite area".

Stages of Community Development

There are numerous papers on the stages of community development: Succession. Nudation, migration, ecesis (adjustment of a plant to a new habitat), competition, reaction and stabilization.

Disturbance

When discussing community development, it implies that there has been some sort of disruption in the vegetation of the community. Grime defined disturbance as "the mechanisms, which limit the plant biomass by causing its partial or total destruction". Disturbance should always be specified according to spatial extent, temporal extent and magnitude.

Type of Disturbance

Grubb summarized information for different types of plants under different disturbances. While he thought it possible to generalize about type of disturbance and plant life form it was difficult to generalize life characteristics and high disturbance levels. For example, demographic characteristics of communities exposed to grazing reveal a high number of perennial plants capable of both sexual and vegetative reproduction. Annuals and biennials are generally located in interstices and fluctuate in relative abundance.

Grazing intensity tends to increase the dominance of clonal plants. The number of species with vegetative reproduction were significantly different under grazing than those with vegetative reproduction. In the same study, seed dispersal was also analysed. Mobile and adhesion, as methods of dispersal, were both significantly different under grazing. Wind dispersal was significantly different under soil disturbance.

On a denuded glacial foreland, pioneer species were plants whose dominant reproductive form was vegetative. Unlike early colonizing species of secondary successions, pioneers of primary succession are usually long-lived perennials that are relatively slow growing.

Research by Vose and White in a burned Ponderosa pine (Pinus ponderosa) community indicated that buried seed did not contribute significantly to seedling recruitment, and seed rain was more important than buried seed. Most regeneration was from seed rain, and the only vegetative reproduction was from the shrub Ceanothus fendleri. It should be noted that this shrub also reproduced by seed. Allogenic disturbances reveal a question to consider: can we take research from autogenic disturbances and extrapolate them to allogenic disturbances?

Intensity and Frequency of Disturbance

Key characteristics suited to intensity of disturbance are very different. Grubb thought special attention should be placed on plants that were able to increase their range after a disturbance and more study was required on life form, phenolgy, and requirements and tolerances.

Vose and White found that intense heat from a pole-stand burn reduced seed rain compared to sites that burned with less intensity. Burning prevented germination of any buried seed. In their study, the authors speculated that post-fire environmental conditions might have been unfavourable to germination and establishment. It is commonly thought that ruderals, with high seed production, generally dominate areas of frequent, high disturbance. Seeds are usually dormant, and these plants are largely opportunistic of the increases in light supply and nutrients in a disturbed site. A summary of data by Grubb in 1985 supported this theory. In natural areas of yearly disturbance, (for example, scour of drift lines by floods), there is a high number of annuals that use the abundant nutrient supply and set seed in the first season. The seeds of these annuals have no specialized means of dispersal and rely on dispersal though the drift material.

Environmental Conditions: where did the Disturbance occur?

Low Temperatures

On the tundra, there are very low reproductive rates and predominance of vegetative proliferation. Most invading species are either self-pollinated, or invest massive resources in vegetative propagules there is a higher representation of species that possess some mode of uniparental reproduction.

Chapin calculated that the carbon cost of producing a tiller of Dupontia fisheri from seed at Pt. Barrow, Alaska was 10,000 times greater that that of producing tiller vegetatively.

A single Betula pubescens in Swedish Lapland produces 40,000,000 seeds thoughout its life, only one of which is necessary to replace the original tree for a stable population.

Are these seeds only important in the colonization of new areas outside the forest or after disturbance such as fire within the forest?

Recruitment from seed is important in open habitats such as fell-fields, but early mortality rates are high. Vegetative reproduction has low recruitment levels but high survival rates due to the physiological interdependence of modules. (fell fields are open communities of scattered individuals aggregated into islands). The plants are dwarfed and prostrate, and the habitat is characterized by little winter snow cover, water shortage, primitive soils and frequent soil movement due to freeze-thaw cycles).Chambers cautioned in diminishing the importance of sexual reproduction in tundra environments. And though compared to temperate climes, sexual role is reduced, for tundra it is still very important.

Callaghan and Emanuelsson indicate that sexual reproduction may be adapted to the exceptional and not too normal circumstances. Environmental conditions that select for sexual reproduction may be very different from those that select for vegetative growth.

Low nutrients

On the tundra and in alpine disturbances, distinct species colonization patterns can be observed. Chambers noted that species with nitrogen-fixing symbionts are frequent colonizers on mineral soils low in nitrogen, including those exposed by retreating glaciers or snowbanks. In alpine tundra areas of severe soil disturbance, where the resultant soils are poorly developed and have low nutrient status, colonization is most frequently from dissemules, mostly seeds dispersed from adjacent communities. In these conditions, it is necessary for plants to produce many seeds that can easily be dispersed

Importance of Soil and Water

Prach and Pysek studied the role of clonal plants in plots that were 10 to 50 years old, grown under different types of habitats. They had acidic emergent bottoms, peaty soils, sandy, reclaimed, abandoned sand pit, ruderal sites ranging from poor to rich and old fields that were wet to xeric. From their observations of old fields, clonal plants dominated under xeric conditions, while non-clonal dominated under mesic conditions. These sites were 50+ years old. This indicates that more long-term studies in old fields are needed under various environmental conditions.

The study they did for plots that were 10 years old concluded the data set was limited and not able to reveal a functional relationship between the participation of clonal plants and habitats.

Stocklin and Baumler found that seeds found in areas around a glacier (primary to secondary successional areas) were extremely dependent on water availability for germination and establishment. Seedling success was dependent on "safe" sites, which were slight indentations or areas covered by dead plant material. Chambers found that in windy, soil exposed areas (tundra/alpine) relationships between soil characteristics and seed attributes significantly affected the spatial distribution of dispersed seeds. This may be one of the primary determinants of species colonization patterns.

Chambers found that with small soil particles, small seeds and seeds with adhesive coats were trapped, but most large seeds moved across the surface; at large particle size, larger seeds were trapped, with larger particles, small seeds moved further downward in the soil as well. In the same study, surface soil temperatures were measured and on dark covered organic soils, the temperature ranged from 35-45 °C on the surface and 20-25 °C at 5 cm depth. On an adjacent site with exposed, loamy sands, midday soil temperatures were about 5 °C lower. Soil temperature can result in higher rates of mineralization and decomposition. Most tundra species had higher germination rates under higher temperatures than what is normally found in the field.

Competition

Small clones spread laterally and are not overtopped by larger clones. Dominance and suppression by non-clonal Genetic variability in plants

Temporal Considerations

Noble and Slatyer, listed the following three plant attributes as most important in vegetation replacement:

- The method of arrival or persistence of the species at the site and after disturbance.
- The ability to establish and grow to maturity in the developing community.
- The time for the species to reach critical life stages.

Epp and Aarssen found interactions between species in recently established vegetation (1 year) are affected more by differences in attributes that are relevant to early life, *i.e* seed weight. This relates to the competitive ability of seeds in establishing themselves in an open area. They also discovered that the older a stand became, other attributes became important in a plant's ability to remain dominant. In a 1 year old plot, lateral spread only accounted for 0.3% of the variation, while in the 11 year old stand, it accounted for 26% of the variation in species diversity. Seed weight (number of seeds was not measured) accounted for 13% of the variability in the 1-year-old plot, but only 0.03% in the 11 year old plot. Non-clonal plants had maximum coverage after 4 years, and the clonal plants had maximum coverage in years 6 and 7. Coverage for the clonal plants decreased after year 7. They also determined a difference between guerilla and phalanx plants. Phalanx had a peak of 70% coverage in year 5 and that decreased to 28% coverage in year 10. Guerilla plants peaked at year 6, at 72% and in year 10 were 61% (this was found to be significantly different).

Species capable of guerilla growth are often more successful in the later stages of succession than plants with phalanx growth; this may be explained by the guerilla's ability to produce widely spaced modules that have a greater chance of penetrating into more or less closed vegetation. Differences in clonal growth forms were also found by Oborny in simulation models of different types of vegetatively reproducing plants. Clonal plants that had a rigid structure decreased in number over time vs spp that had plastic growth form.

Vegetative reproduction becomes of more importance in later stages, when dense cover and a thick litter layer limit establishment from seed. But there are limitations to vegetative reproduction as well. Belsy found that populations of Sporobolus fimbriatus and Pennisetum mexianum began to replace Digitaria maroblephara and Sporobolu ioclados after 5 years of non-grazing. This is due to the fact that S. fimbriatus and P. mexianum are both tall rhizomatous species, whereas Digitari maroblephara and Sporobolus ioclados are both stoloniferous. The build up of the litter layer suspended the stolons above the ground. This was also true for species that reproduced by seed. The seeds could not germinate with the heavy litter layer.

Factors that Influence Colonization: Specifically in Regards to Sexual and Vegeta Reproduction

It has been hypothesized that the role of clonal dispersal in succession is evident both in early stages, due to the rapid capture of space, and in late-successional stages, due to the rapid filling of gaps in more or less closed vegetation cover. In essentially closed cover, seeds do not have the opportunity to germinate, nor do seedlings have the chance to establish. Competition for nutrients, water and light is just too great. This would imply that eventually all plant communities should be dominated by plants that rely heavily on vegetative reproduction. Why is it then, that there are many plant communities that appear to stabilize with a combination of plants that use both seed (sexual) and vegetative (asexual) reproduction? Research on the role of clonal species is lacking, and an understanding of their behaviour is based to a large extent on speculation. The factors that inhibit or promote one reproductive strategy over another are not well understood.

CONCLUSIONS

The direct genetic consequences of self-fertilization and asexual reproduction are quite different. Self-fertilization causes progeny to be heterozygous at only half as many loci, on average, as their parents, and highly selfing populations are composed primarily of homozyous genotypes. Asexual reproduction, however, results in progeny that are genetically identical to their parents, barring somatic mutation, and asexual populations therefore may be highly heterozygous. Because self-fertilization and asexual reproduction both prevent exchange of genetic material among family lineages, however, the number of genotypes found in populations of predominant selfers or obligate asexuals is usually much smaller than would be found in a population of sexual outcrossers with the same allele frequencies. For similar reasons, surveys have repeatedly shown that a greater proportion of the genetic diversity found in selfing or asexual species is a result of differences among populations than in sexual outcrossers (Brown, 1979; Gottlieb, 1981; Hamrick and Godt, 1989, 1996). In short, the great diversity of reproductive systems exhibited by vascular plants is matched by a similar diversity of genetic structures within and among their populations. Indeed, the diversity of reproductive systems may be the predominant cause of the diversity in genetic structure.

Differences in genetic structure associated with differences in reproductive systems were once commonly invoked as evolutionary forces governing their origin (Stebbins, 1957; Grant, 1958; Baker, 1959). Although such differences may help us to understand why some lineages persist and diversify and others do not, we now realize that to understand the origin of alternative reproductive systems we must look for benefits and costs associated with individual survival and reproduction. Moreover, the comparison of self-fertilization and agamospermy shows that the most important distinction for hermaphroditic organisms is between uniparental and biparental forms of reproduction. Uniparental reproduction, whether through selfing, agamospermy, or apogamy, excludes sexual outcrossers from contributing to some offspring that will form the next generation. If selfers, agamosperms, or apogams are able to contribute to some sexual offspring, genotypes promoting that mode of reproduction will be over-represented in the next generation, reflecting the cost of outcrossing.

To the segregational cost of outcrossing can be added another: selfers, agamosperms, and apogams are able to produce offspring even under conditions that prevent the union of gametes produced by different individuals. The benefit of this reproductive assurance seems so apparent that it is surprising how few experimental studies have been done to document it and how equivocal their results are (Eckert and Schaefer, 1998). When plants reproducing uniparentally benefit from reproductive assurance or are over-represented in the next generation as a result of donating gametes to the outcrossed progeny of other individuals, they eventually will replace sexual outcrossers in the population, unless the progeny of sexual outcrossers are substantially more likely to survive and reproduce. Thus, the relative fitness of different types of individuals competing in a population and the frequency with which different types are formed will determine whether outcrossers persist in the short term or are replaced by selfers or asexuals.

6

Transgenic Plants for Recombinant Proteins

Numerous heterologous recombinant proteins of economic value have been produced in transgenic plants. These proteins are often produced and accumulate in plant organs like leaves, fruits, roots, tubers and seeds. Also these proteins are targeted to different sub-cellular compartments, such as cytoplasm, endoplasmic reticulum (ER) or vacuoles. However, extraction and purification of proteins from biochemically complex tissues is a laborious and expensive process, which becomes a major barrier to large scale production of these proteins in transgenic plants. Even the in vitro systems for manufacture of recombinant proteins can be slow, low yielding and expensive. In view of this, a plant system based on natural secretion from the roots of the intact plants has been suggested and tried.

The plant roots have developed a sophisticated mechanism involving secretion of different chemicals into the rhizosphere. In a study, reported in 1999, three heterologous genes (genes for green fluorescent protein or GFP from jellyfish, Aequore victoria, human placental secreted alkaline phosphatase or SEAP and xylanase from bacterium, Clostridium thermocellum) were used for the production of transgenic plants, demonstrating that root secretion (termed rhizosecretion) can be successfully exploited for the production of recombinant proteins.

Transgenic Plants for Value Added Speciality Crops

There are many examples of chemicals that have already been produced using transgenic plants as production systems, although they may not have been commercially exploited in all cases.

LEGITIMATE CONCERNS ABOUT TRANSGENIC AGRICULTURE

There are problems with genetic engineering. As an engineer myself, even though I work with electronics instead of genes, I am naturally disposed to be sympathetic to genetic engineering, but that doesn't mean that it should be practiced without a concern for its dangers.

So the next part of this report is devoted to a survey of some of the legitimate concerns about genetic engineering of crops. Later we will mention some other concerns that are not realistic at all.

Monkeying with Mother Nature

Some people think that this is an enterprise that should be left to God or to Mother Nature, that man was never intended to monkey around with other species' genes. I respect this point of view, even though I don't agree with it. But it can't be the basis of an argument. Whoever claims to know what God intends usually can't prove it, and can't be talked out of it.

Some people have religious or ethical concerns. They might point to Leviticus 19:19, which prohibits crossbreeding. Vegetarians may reasonably decide that their food should not contain genes derived from animals.

Jews and Muslims may reasonably decide that their food must not contain any genes derived from a pig. Some religious scholars believe that a gene loses its identity when it is copied and the copy is inserted into a target species. That point of view would remove some, but not all, of the religious objections to genetically modified plants. Other advances in biotechnology have drawn most of the attention of clerics and ethicists. These include cloning, organ transplantation, research using foetal tissue, etc.

Food Safety

Crops modified in any way might not be safe to eat, so any major change in the food supply should be tested. This applies to changes made by genetic engineering but it ought logically to apply even more to changes made by other techniques. To a great extent, genetic engineers know what they are doing. There can be unanticipated consequences, but by comparison, all other methods of improving crops involve an element of luck. The conservative approach is to test all crops whose genetics has been modified in any significant way.

An example of a possible safety issue was brought out clearly several years ago. Although soybeans are a good source of protein, soy protein is low quality. It doesn't have enough of the essential amino acid methionine. So scientists in Nebraska planned to transfer a gene from a Brazil nut to a soybean to get better quality protein from soybeans for use as an animal feed. Unfortunately, some people are allergic to Brazil nuts and it turned out that the better quality protein was one of the Brazil nut allergens. Since this fact was quickly revealed by testing, the genetic modification project was abandoned. This example shows that testing for safety is necessary. It also shows that such testing is being done and is working. But can new foods ever be tested enough for complete assurance of safety?

Another way to develop crops with new traits is to cause random gene mutations and select for them. This is analogous to spraying on the pages of a recipe book, then following the distorted recipes to see which work well, and recopying those recipes. Breeders have induced mutations using radiation, chemicals and high temperatures.

Since the effect of mutation is random, it makes sense that crops developed by mutation ought to be even more thoroughly tested than crops developed by genetic engineers since the genetic engineers are not relying on luck to get their improved traits. Yet there is essentially no regulatory process for plant breeding in the United States, although Canada requires that all new types of cultivars be tested.

Even conventional breeding techniques can accidentally create harmful foods. In a famous example, an improved variety of celery caused farm workers who picked the celery to become hypersensitive to sunlight. In another example a potato variety, Lenape, was withdrawn from the U.S. market in the 1960s when it was found to contain dangerously high levels of potato toxins (solanidine glycosides).

Even without mutations, there is a large pool of genetic variability in every variety or species. This means that unfavorable combinations are possible. In every instance of sexual reproduction the child gets some genes from each parent, in a random assortment. If John and Jane have a few hundred different genes (and about 30,000 that are identical), their children will each inherit a different subset of John's genes and a different subset of Jane's genes. Nobody can predict the characteristics each child will inherit from its parents. Sometimes, two apparently healthy parents have a child with a genetic disease. Similarly, sometimes two plants which bear nutritious food can have offspring which are more toxic. This is not an argument against having children or against breeding crops, so it ought not to be an argument against transferring genes by biotechnology.

Conditions of growth can also affect food properties. Certain inconspicuous fungi can turn a wholesome food into a poisonous food. Every year there are deaths from *ergot*, a fungus that infects wheat and rye, and from aflatoxin, caused by a mold that infects peanuts and corn.

In summary, genetic engineered crops need to be tested for safety. In the US, transgenic crops are tested much more strictly than crops developed by traditional breeding. So far the testing that has been carried out has been sufficient to protect the public. During the ten years that we have been eating transgenic foods, nobody has ever been exposed to unsafe genetic engineered food. Meanwhile there have been many thousands of deaths because of unsafe conventional food. So it seems to me that the issues of food safety are being better managed for genetic engineered foods than for conventional foods.

Environmental Concerns

The third thread of concern is for the wild environment. Suppose a gene from an unrelated species is transferred to a crop species and then the modified crop produces pollen which fertilizes a wild plant. Or suppose some of the crop's seeds are carried by birds or by wind into the wild. The wild plant could reproduce and the gene could become fixed in the wild population. If it conferred an advantage, a wild plant that had been barely making it in the struggle for existence could turn into a dominant species. There are many examples of plants taking over an environment. Usually they are natural plants introduced from a distant continent. In the American south, kudzu is a decorative plant that escaped and is spreading out of control. In the northeast we see the same thing happening in marshes, being taken over by purple loosestrife. Pasture land in the American west is being invaded by cheat grass. These plants have no natural enemies and can overrun an ecology and devastate it.

So suppose a genetic engineered crop has been given a gene which makes it hardier. Suppose it gives the plant a tolerance to salty soil, or to cold, or to dryness. It is reasonable to fear that if the crop's pollen fertilizes a wild relative, that relative could produce a race of super weeds.

The solution to this concern is, again, testing. Scientists must study the plants growing wild in the area, determine which are closely related to the modified crop, experiment to see if hybridization is possible, and require that the crop be grown only in conditions for which hybridization is very unlikely. Or else, determine that the trait, in the wild relative, will not matter much.

Sometimes this is fairly easy. You can be pretty sure that soybeans will not hybridize with wild relatives because they self-pollinate and because the wild relatives live only in Asia. Corn can only hybridize with its wild relative, teosinte, found only in Guatemala and southern Mexico. Sugar beets are harvested before they produce flowers (unless they are being grown for seed) so they cannot pollinate other varieties. But for some other crops the testing should be much more extensive and in some cases it will not be allowable to grow the genetically modified crop in localities where wild close relatives are found.

Earlier we mentioned a variety of corn which could grow well in soils with high levels of aluminum. That variety was developed in Mexico, but cannot be field-tested there, perhaps because of the fear that it could pollinate its relative, teosinte, and give teosinte an advantage over other local wild plants. If careful analysis confirms this danger, there might still be a way around the problem. It is now possible to produce plants which are male-sterile. They produce no pollen, or only defective pollen. A farmer could then plant conventional corn as a pollen source and aluminum tolerant corn for his main crop. Since the plant with the extra gene is infertile, it would not be able to spread its gene by pollination. This still leaves the possibility that a seed could escape, carrying the special gene, but corn cannot live at all as a wild plant—it cannot reseed itself—so this avenue of gene transmission is much less likely.

Transgenic salmon have been engineered to grow faster than their wild relatives. There is a concern that they might escape from confinement pens and reproduce, or even cross with their wild relatives. Nobody can confidently predict the ecological effect of this. The transgenics could monopolize the wild salmon's food supply or be preferred as mates. If they were effective at attracting mates but less prolific breeders, salmon populations could crash. To prevent all these possibilities, Aqua Bounty Inc., the company developing transgenic salmon, plans to use only sterile females in commercial production.

Not every species that escapes into the wild will be a problem. Most crops will simply die out because they can't compete with hardier wild plants. In one experiment, rapeseed plants, both transgenic and conventional, were grown in a field but never harvested. Scientists then followed the subsequent history of the field for ten years. All the crops declined in numbers from year to year. After the fifth year, none of the genetically modified crops could be found at all, and after ten years there were only a few crop plants of any type remaining in the field.

In a more colorful example, during the nineteenth century, a wealthy and eccentric man brought to the United States populations of each type of bird mentioned in the works of Shakespeare. Only one species was able to establish itself. That species, however, was the starling, now found in large numbers in every part of the United States.

Some species out of their natural place can enrich the environment. The European honey bee (Apis mellifera) was introduced to America in colonial times. Besides its value as a honey maker, it is the principal agent of pollination for many staple crops, which are also European imports.

Organic farmers have a different concern. They consider a genetic engineered crop to be automatically *non-organic* even if it is grown without pesticides or chemical fertilizers. They have expressed the concern that their crops might be cross fertilized by pollen from a gene modified crop. Organic farmers have every right to be protected from this problem. It is no different, in principle, from the problem faced by the seed companies who grow seed for sale. It is solved partly by keeping the different crops far apart, and partly by more active techniques, like barriers.

There is another concern. Suppose a crop is developed which is resistant to a certain insect or fungus. Evolution is like an arms race. Insects or fungi can evolve to overcome whatever defence has been built into the crop. For example, cotton with the Bacillus thuringiensis toxin will eventually lead to insects evolving a resistance to Bt toxin. But Bt is used by organic growers to control certain insects. I've used it myself. If boll budworms evolve Bt resistance, organic cotton farmers will not have alternative controls. Non-organic farmers might turn to some other insecticide, but even they would like some strategy to delay the evolution of resistant insects.

The solution to this problem is the so-called refuge strategy. Instead of growing only Bt cotton in a large field, the farmer must grow a mixture of Bt cotton and conventional cotton. This is an EPA rule. The theory is that then the insect with a lucky mutation who can tolerate the Bt toxin will have no advantage over the other insects without the mutation so Darwinian selection will not tend to increase the numbers of such insects in a population. This strategy is not simple. What percentage of the cotton in a field must be conventional, and how must the two types of plants be spaced? What about how the field is used in the following season? Personally I am not enthusiastic about the refuge strategy, but so far it has worked as advertised. Yet, as more and more acres are sown with Bt crops, year after year, it seems as if the insects must eventually evolve the resistance.

Eventually genetic engineers will develop better ways to delay the evolution of insect resistance. Many plants have natural insect defences which they use only when they are being attacked. Today's Bt crops express their toxin all the time, which gives their insect adversaries a constant environment in which to evolve. It would be much harder for insects to evolve a resistance to a varying environment. So it would be better to control the gene for Bt toxin selectively. For example, if scientists could identify a control gene that turns on when the plant is attacked, they could use an an identical copy of that control gene to turn on the Bt toxin gene only when it was needed. Even easier, a control gene could be used which would turn on the toxin gene in response to a cheap and harmless chemical which the farmer would spray only when deemed necessary.

The problem of evolved resistance is not new to genetic engineered crops. Insects also evolve resistance to chemical pesticides and to other control methods. Some farmers used to protect corn from rootworms (beetles) by growing corn in a field used every other year for soybeans. The corn rootworms soon evolved the habit of laying their eggs in soybean fields. Farmers and agricultural scientists are unlikely to ever find a perpetual solution to suppressing insect pests. One widely voiced concern about transgenic crops is quite ridiculous, the fear that they could transfer unwanted genes to the bacteria that live in our intestines. Many genetic engineered crops contain a gene for resistance to an antibiotic, such as kanamycin, which was transferred along with the useful genes. The reason is that most genetic transfer attempts are unsuccessful in some way so the genetic engineers try to transfer the useful genes into hundreds of plant cells at once, hoping that a few cells will be successfully transformed. But after a gene has been transferred into a plant cell, it takes much hard work to create a whole mature plant from that single cell and the genetic engineers want to avoid that work except for the cells that have been transformed. But it is hard to tell when a gene transfer has worked if the gene only functions in the mature plant. The antibiotic resistance gene acts as a marker. When the transformed cells are treated with the antibiotic they survive, but cells that were not transformed die. Only the marked cells are turned into mature plants for further experimentation.

But the concern has been voiced that E. coli bacteria which live in our intestines could obtain these genes and become antibiotic resistant. We certainly don't want that to happen, at least not by accident.

We can't transfer genes by eating them, but, unfortunately, bacteria *can* take up DNA from their environment and incorporate it in their genome, although this phenomenon is extremely rare. It is billions of times more likely for bacteria to acquire a gene for antibiotic resistance by natural mutation. Each of us has a few such bacteria in our intestines now. That's where the marker genes originally came from. If we are not taking the antibiotic medicine, the percentage of these bacteria will be quite negligible. The only defence against both the natural evolution of antibiotic resistance and the circuitous route through transgenic crops is to minimize the use of antibiotics.

Nevertheless, although the best scientific evidence is that antibiotic resistance genes would not be a problem, genetic engineers have stopped using them as marker genes. The alternative marker gene now favored confers resistance to a herbicide instead of resistance to an antibiotic. Another kind of marker gene comes from a bioluminescent jellyfish. It promotes synthesis of a fluorescent protein. When this protein is exposed to ultraviolet light, it produces visible green light, clearly indicating that the desired gene has been transferred to the target organism. Also, after a genetic engineer has created a useful new variety of plant with both useful genes and marker genes it is possible to eliminate the marker genes from future generations of the plant using conventional cross-breeding.

Economic and Social Concerns

There is one last concern often expressed. Although the gene transfer technology is available worldwide, some people worry that a few large companies would get control of world agriculture, and further, that small farmers in the poorer countries would be at an ever increasing disadvantage as their competition becomes ever more productive. The counter-argument for the first concern is that we already have antitrust laws in place. The counter argument for the second concern is that the poor third-world farmers could also adopt more efficient farming practices. The experience of the last thirty years, the so-called green revolution for which Norman Borlaug received his Nobel Peace prize, is that third world farmers can and do adopt new technologies. However, genetic engineering is yet one more technology which is making agriculture more dependent on large companies.

A Debt to Critics

The two most serious concerns about transgenic agriculture are food safety and environmental impact. So far the record of the technology has been enviable. There have been no documented cases of any illness or any environmental damage.

For this the scientists developing the technology must and do owe a debt of gratitude to the people who have raised doubts. Responsible critics have suggested problems and the scientists have been able to take appropriate precautions, or have cancelled dangerous experiments. The anticipated mishaps didn't happen because they were anticipated.

For example, if no critic had raised the possibility of allergies, would transgenic foods be tested for known or likely allergens? If no critic had raised the possibility of insects evolving resistance to Bacillus thuringiensis, would the Bt cotton be grown with non-transgenic cotton close by? Would transgenic salmon farming be limited to sterile females if no critic had raised the possibility of escape and crossbreeding with wild stocks? It takes no credit away from the scientists to acknowledge that the enviable safety record of genetic engineering in agriculture derives as much from its critics as from its inventors.

Part II

New technology, proven to deliver advantages to farmer, consumer, and the environment but that there are reasons to be concerned because, like any new technology, it could be misused. Since the US has been the leader in adopting genetic engineering for agriculture, our government agencies have developed some standards for assuring food safety and environmental safety. Any genetic engineered product must meet these standards before it can be grown commercially.

The Moratorium/Ethics of Genetic Researchers

We would like to mention something of the history of this research. During the 1970's, without any government regulation whatsoever, all the researchers in the field of genetic engineering adopted a self-imposed moratorium on further research for one year. They spent that year in developing and agreeing to a set of standards for experimental work to assure that the public would be protected from danger. To the best of my knowledge this is the only example of its kind in the history of technology.

At the beginning of the public funding of the human genome project, it was the scientists, not the politicians, who decided to devote five per cent of the funding to a study of its legal, ethical and social implications.

These events show that concerns for safety and for the social consequences of their research were on the minds of genetic engineers from the beginnings of their field, and that they have, as a group, exceptional ethics.

A Movement to Frustrate Transgenic Agriculture

Although there are legitimate reasons to oppose genetic engineered agriculture, or at least to demand the most careful controls, there are a community of opponents who have taken their opposition beyond what is ethical. I am not talking about a reasonable opposition expressing the concerns summarized earlier, but rather about an opposition for which the ends justify the means, including lies, vandalism, etc. I need to stress this point. There are many people, sincerely opposed to genetic engineered crops, whose ethics I do not question. I believe that most of the opponents of biotechnology would fall into that category. Yet, in too many cases, those sincere concerns are based entirely on misinformation which originated in deliberate lies and fear mongering. We are here exposing the ethics of the people who have created the lies. These opponents must have some motive, and it seems that an alliance has emerged between at least four groups, each with its own agenda.

First there are people who sincerely believe that genetic engineering is ethically wrong, and that anything they do to stop it from happening is therefore right.

Second there are foreign governments and their constituencies who are worried about American domination of agriculture. Closely related to this are advocates of organic agriculture who seem to be engendering public fear to make their own products more salable. Third, there are environmental groups who have been misled by a radical fringe and have become willing to do anything to stop genetic engineering agriculture. The most conspicuous among these groups is Greenpeace.

Finally, there are people opposed to capitalism or to large businesses dominating agriculture.

Opposition from environmental groups is particularly frustrating to me. Most measures which benefit the environment require people to give up something, and they don't like to do that. Recycling is nearly painless and saves money, but many people won't make the effort. A few degrees adjustment of a thermostat could save vast amounts of energy, but most people would rather be comfortable. We end up settling for half measures. But here is a technology that benefits the environment without asking people to give up anything, and its biggest opposition comes from environmental groups.

MISINFORMATION ABOUT FOOD SAFETY

There is a deliberate campaign to frighten people about the safety of the food supply. This campaign has worked successfully in England and in much of Europe.

Dr. Arpad Pusztai, who worked at the Rowett Institute in Aberdeen, Scotland, performed an experiment. It began when a gene was transferred from a poisonous plant, the snowdrop, into a potato. The transferred gene specifies the production of a poisonous compound called lectin. Dr. Pusztai proceeded to experiment with rats. Some rats were fed with the raw potatoes which were genetically engineered to contain the poison. The rats in the control group were fed ordinary raw potatoes and were also given the amount of lectin poison which the first group of rats would have gotten from eating the transgenic potatoes. Both groups of rats developed malformed organs, and there was no statistically significant difference between the rats who consumed the poisonous potatoes and those who consumed the poison.

However, Dr. Pusztai claimed that his data showed that the rats who ate the genetically modified potatoes had more deformed organs. No scientific journal would publish Dr. Pusztai's interpretation, and his institution would not support him. He hired an independent statistician to review his data, who also considered the data to show no difference between the two groups. But Dr. Puzstai held to his opinion. Eventually the disagreement became serious enough that his connection with the Rowett Institute was ended.

The opponents of genetic engineering, mostly in England, have blown this result into a cause celebre. Dr. Pusztai is portrayed as muzzled by the scientific establishment, although the British medical journal Lancet eventually published Pusztai's paper over the recommendations of its reviewers because of the widespread public interest. The British tabloid press covers this story continuously, with lurid photographs of deformed rat organs. The potatoes genetically engineered to be poisonous became synonymous with all transgenic food, called Frankenstein food in the tabloids.

There are varieties of potatoes, bred to be eaten, engineered to resist insects, viruses and fungi. All these varieties have been fed to rats and have never harmed them. The opponents of transgenic food have no explanation for that-they are content to use one probably misinterpreted experiment with potatoes nobody will ever eat, to stir up doubts about food safety.

During safety testing of transgenic tomatoes, Dr. Belinda Martineau discovered that when rats are fed huge amounts of tomato paste they can develop stomach lesions. It doesn't matter whether the tomatoes are transgenic or not. But before there were transgenic tomatoes, nobody had ever fed rats such large doses. Martineau's results were reported to the FDA but opponents routinely call this a "cover-up" of a health hazard with transgenic tomatoes.

As another example, I read an op-ed article in the Boston Globe in August of 1999, written by Paul Billings, a member of the board of the Council for Responsible Genetics. The gist of the article is that dangerous untested foods are being foisted upon an unsuspecting American public, by mad scientists. Every genetically engineered crop has been tested for safety. The testing has been much more extensive than that for any other foods, including foods developed by radiation induced mutation. Dr. Billings knows this. He knows about the government testing rules that establish the safety of each individual crop. He wants genetically modified food to be tested as strictly as drugs are tested.

Billings' op-ed article contains only one 'fact' that most people would not have known before-the rest is either his opinion or just wrong. He says that transgenic soybeans have been shown to be deficient in a certain unidentified nutrient. It is not easy to track down the source of this 'fact', but I did it. The nutrient in question is a phyto-estrogen (also known as a phytosterol). Although phyto-estrogens are not essential to human health, there is some indication that they help prevent cancer. The study that indicates that transgenic soybeans are deficient in phyto-estrogens comes from Dr. Marc Lappe, who wrote the book "Against The Grain", a polemic against genetic engineering and especially against the Monsanto Co., the leading company in the field, which developed the soybeans in question.

Here is how Dr. Lappe established that genetic engineered soybeans are defective. Understand that there are dozens of varieties of soybeans with the herbicide tolerance trait and over a hundred varieties of conventional soybeans. Dr. Lappe compared one conventional variety with one transgenic variety. He found a 12% difference. But individual soybean varieties vary by more than 100% in their phyto-estrogen content. Also phyto-estrogen content is not stable. It declines with storage, by much more than the 12% difference. But Lappe at least reports his data along with his biased interpretation of it. Dr. Billings reports only the interpretation without any indication that it comes from a biased scientist whose own data show insignificant variations in a nutrient whose role in human health is not even firmly established. Dr.

Billings seeks only to mislead people. Not one reader in a thousand would do what I did, track down the data. His purpose is to plant a little seed of doubt about food safety, hoping that it will fester in our minds, mingle with similar misinformation, and eventually become accepted fact. Not everything misleading is false. Propagandists are very skilled at making true statements that seem to imply something quite different. For example, they frequently say that the FDA's rules about testing genetically modified food are voluntary, implying that some testing doesn't get done. Under current law, FDA has no authority to require safety tests for any food, transgenic or conventional, although it can prevent the sale of foods it considers unsafe. But, in fact, each developer of a transgenic crop has consulted with FDA and performed every test FDA suggested. They also like to say that safety test results are trade secrets. True, it would be legal to keep test results secret. But no developer has done it.

Earlier I mentioned a problem with a soybean with a Brazil nut gene. People allergic to Brazil nuts should not expect to have to avoid soybeans, but the allergen was identified by testing and therefore the modified soybeans were never created. Also, the scientists who demonstrated this published their results in a scientific journal, and this led the FDA to not allow any gene to be transferred into a food from a species known to cause allergies. That should be seen as evidence that the genetic engineers are responsible people, and that testing is working well. But the unethical opponents of biotechnology routinely present this episode, carefully worded, as if there was a near disaster, revealing gross problems with the current regulatory system.

Tryptophan Deaths

One of the most active anti-transgenic groups is *Mothers For Natural Law* which spreads the following half-truth-that, in 1989, 37 people died and thousands were paralyzed by consuming tryptophan made by genetic engineered bacteria. Half-truth because there were deaths and illnesses (eosinophilia myalgia syndrome) caused by tryptophan, sold by the "health food" industry.

Tryptophan is one of the twenty amino acids which are needed by every living thing. All bacteria already contain genes to make tryptophan and the tryptophan sold as a dietary supplement is made industrially using bacteria and is then purified. Showa Demko Ltd., the company whose tryptophan caused illnesses, genetically engineered bacteria to make *more* tryptophan than was needed for the bacteria's own life cycle. The illnesses had nothing to do with the genetic engineering. Some cases of eosinophilia myalgia syndrome were traced back to Showa Demko's tryptophan manufactured as far back as 1983, years before the company used genetic engineered bacteria. It is now known that eosinophilia myalgia syndrome is caused by consuming excessively large doses of tryptophan, from whatever source. Back in 1989 it was thought that the cases of eosinophilia myalgia syndrome had been caused by some kind of contamination. All tryptophan molecules made by living things are chemically identical and there is no way that tryptophan made by genetically engineered bacteria could be different from tryptophan made by any other bacteria.

The tragic epidemic of eosinophilia myalgia syndrome makes a very good argument for more scrutiny of the health food industry. Presenting it as an indictment of transgenic food is a huge distortion.

It's Unlike Anything in Nature

Advocates of genetic modification of crops often say that it is not significantly different from ordinary breeding techniques. They say that virtually every crop is genetically modified and that people have been genetically modifying plants and animals for several thousand years. This is, of course, true, but are the genetic transformations now possible through biotechnology different from classical breeding in some fundamental way?

The opponents say that the new gene transfer techniques are completely different from anything that nature has ever allowed. Since this is only a matter of how the two sides define *fundamental,* it really isn't a case that illustrates an unethical behaviour by either side. There is one minor exception.

The opponents like to illustrate their case by pointing to a tomato with a gene from a fish. This example seems to be selected from all the myriad possibilities because it strikes a chord of negative emotion. We just don't think that anything from a fish belongs in a tomato. This poster child for the opponents is played up endlessly. Their flyers and posters show a tomato with fins, or sometimes a whole fish with a stem and a few leaves. Sometimes the tomato is a strawberry. One is supposed to think that these are typical examples of genetic engineering. They are not!

It is possible to transfer a gene from a fish to a tomato plant. It was tried by DNA Plant Technology of Oakland, California. The fish, an arctic flounder, can tolerate very cold water because its blood contains a natural antifreeze. The hope was that a tomato plant would also be cold tolerant. When the resulting plant was tested, it was a failure. The company abandoned the project and has no plan to try again. All the posters portray is a just-so-story, a product that doesn't exist. In fact, no plant product on the market today contains a gene from any kind of animal, with one exception-there is a gene from a luminescent jellyfish used as a marker, an indication that the gene transfer has been successful. (Images courtesy of Steven Haddock, Monterrey Bay Aquarium Research Institute)

But the story still bothers people even when they know it doesn't have much to do with anything we already eat. There is a feeling that it somehow goes against nature to make such huge changes in an organism's genes.

It might be useful to examine a few cases from nature, which can be more complex than most of us imagine. At the very least, it will be interesting. There are natural examples of genetic engineering, and they are actually quite close to our lives.

Wheat is sometimes called the staff of life. Yet wheat has a complex genetic story. It is the result of three separate instances of natural genetic engineering. To introduce these changes, we need to explain that wild grasses similar to wheat have their genes dispersed among seven pairs of chromosomes. One of the earliest known domestic wheat varieties is einkorn wheat (Triticum monococcum), which has seven chromosome pairs, like a wild grass. But another variety of wheat, emmer wheat, has 14 chromosome pairs. It resulted from an "impossible" cross species mating with another wild grass (Aegilops speltoids). This cross preceded modern biotechnology by several thousand years. It either happened by itself or with the help of a Sumerian farmer. This new plant had new characteristics that breeders, with ordinary breeding and selection, exploited to produce many modern varieties, such as durum wheat, which has grains that are easy to separate from the hulls. But natural genetic engineering was not finished with wheat. Around the time of the Roman empire there was another "impossible" cross species mating with a third wild grass (Triticum tauscii). The resulting new variety of wheat, bread wheat (Triticum aestivum), has twenty one chromosome pairs, the complete genomes of three separate species of grasses. This last mating brought in the genetic recipe for gluten, which makes dough springy and lets it hold together when yeast makes it rise.

The record of these crosses is written in the genomes of wheat varieties and in analyses of grain from archaeological sites. But the latest step in the series is a grain plant with twenty eight chromosome pairs. It is the result of a wheat-rye cross that happened with human help only very recently, but which made no use of the new gene transfer technology. Wheat has been involved in three "impossible" cross species matings during its history as a human food, none relying on the modern DNA technology.

But the DNA manipulation techniques themselves rely on methods developed by nature. To cut a DNA molecule at a specific place, the genetic engineers rely on a collection of natural enzymes called restriction enzymes, each of which recognizes a specific site to make its cut. To join two pieces of

DNA, they rely on a natural enzyme called DNA ligase. To copy DNA they rely on the natural enzyme DNA polymerase. These enzymes are used by living cells to manipulate their DNA. In a few instances, they are even used to manipulate another creature's DNA.

There is a species of bacteria, Agrobacter tumafaciens, whose way of life is to invade a plant and cause it to create a gall, a home for the bacterium. It works its will on the plant by invading its cells and stitching a few of its own genes into the plant's DNA. In effect, Agrobacter tumafaciens is a natural genetic engineer who changes the genome of the infected plant so that it produces food and protection for the bacterium. How different is that from what human genetic engineers do? In fact, one of our ways to transfer a gene into a plant cell is to use A. tumafaciens as a "vector".

There is a fat green caterpillar, the tomato hornworm, that eats tomato plants, and there is a parasitic wasp that lays its eggs in the body of the caterpillar. The wasp larvae use the live caterpillar for food. But why doesn't the caterpillar's immune system attack the wasp larvae? Because the wasp has evolved a partnership with a virus. The virus is carried by the wasp into the caterpillar, where it goes to work changing the caterpillar's DNA, modifying the caterpillar's immune system to the benefit of the wasp larvae and the virus. The photo below shows a tomato hornworm covered with the cocoons of the parasitic wasps that grew as larvae within its body.

We need to look no further than our own bodies for a very ancient example of a cross species mating. Within each of our cells there are tiny bodies called mitochondria, which produce the cells' energy. Each mitochondrion is the descendent of what must once have been a free living bacterium. The mitochondria have their own DNA and they make their own enzymes. In fact, they would have all the machinery needed to run a cell, except that they, eons ago, transferred most of their genes into our nuclear genome.

These and other examples of natural DNA mixing across species and even between plants, animals, bacteria and viruses, show that nature invented genetic engineering before mankind did. Nature even goes to the exact opposite extreme. There is a species of fish that cannot reproduce except by a cross species mating. The Amazon molly (Poecilia formosa), a tiny fish just a few inches long, is a species with no males. Every Amazon molly is a female. It bears its young alive, like its better known relative, the sailfin molly (Poecilia latipinna), which is commonly kept in home aquaria. How can a fish reproduce with no males? The Amazon molly borrows the services of a male sailfin molly. She mates and the sailfin's sperm enter her eggs, causing them to begin development. But the male sailfin molly makes no genetic contribution to the developing embryo. The DNA in his sperm is wasted, which is why we can consider the Amazon molly a totally different species. There are numerous other species all across the animal kingdom which have dispensed entirely with males, and reproduce by parthenogenesis, but it is certainly a surprise to find a species which relies on males of another species to fertilize its eggs. But not so much of a surprise as to learn of a species of cypress in North Africa (Cupressus dupreziana) which plays the trick in reverse. The pollen of the cypress requires the female parts of a different tree to produce its seed cones. The structural and nutritional parts of the seed cones are built by the female, but the genetic component of the seeds comes entirely from the pollen. (In medieval times, it was supposed that humans reproduced in this way, with all heredity carried by the sperm while the mother provided only nutrition and living space for the growing child.)

It is true that modern methods can speed up the processes which transfer genes between species, genera, families, even kingdoms, by millions of times and channel them into directions of our own choosing. Ultimately what we consider to be natural is a personal decision. But that decision should not be affected by street theater. Nature can give you examples of almost anything you can imagine.

SOME EARLY FRUITS OF TRANSGENIC AGRICULTURE

Rice with Vitamin A

Rice does not contain very much vitamin A. In the poorer parts of Asia, where rice is almost the only food of the rural population, a vitamin A deficiency is common, leading to early blindness. Now Drs. Ingo Potrykus and Peter Beyer, two genetic engineers, have transferred the genes for vitamin A from other species into rice, creating a strain of rice which is rich in vitamin A—the amount of rice in a typical third world diet could provide about fifteen per cent of the recommended daily allowance of vitamin A, sufficient to prevent blindness. Now that a few plants with this trait have been created, they are being cross bred with other varieties of rice using conventional breeding techniques, as has been done for centuries. Such cross breeding could further increase the vitamin A content.

The development of rice with vitamin A was carried out at the Swiss Federal Institute of Technology, making free use of patented technology and of the earlier research which had established the basic facts about how plants synthesize vitamins.

The corporations holding the various patents all agreed to cost-free use of their patents as long as the rice was to be provided free to poor third-world farmers. The new rice strain was then turned over to the International Rice Research Institute, a non-profit organization based in the Philippines, where it will be evaluated for its adaptability to various growing conditions, food safety, and environmental impacts, etc. The IRRI preserves thousands of varieties of rice with different genetic characteristics, so the new strain can be cross bred to produce varieties suitable for almost any locality. The result is that rural Asians can soon expect to retain normal eyesight. Genetic engineers also intend to produce a rice variety rich in iron, because iron-deficiency anemia is a common problem in the same rural populations. But this is a more difficult problem than increasing rice's vitamin A content. Rice contains a substance called phytate. Phytate prevents the body from absorbing iron, so it does little good to breed for increased iron content, and the rice plant cannot reproduce without adequate phytate in the grains. Dr. Potrykus hopes to be able to find a gene coding for a protein that will break down phytate when the rice is cooked.

No-Till Agriculture

The world's biggest environmental problem is loss of topsoil to wind and drainage. The US experienced its dust bowl during the early part of the 20th century as a result of ploughing up the prairie. The problem is much worse in tropical soils, which may have a thin, inches thin, layer of topsoil above a type of soil which, when ploughed, turns into a non-porous concrete-like substance. One field, one crop, once. The problem, both in the prairie and in the tropics, is deep ploughing, which kills weeds which would otherwise crowd out the desired crop.

The solution is called low-till agriculture. The soil is broken up but not deeply ploughed. Weeds are killed instead by herbicides. A herbicide of choice should be cheap, quickly biodegradable and non-toxic. An excellent choice is a chemical called glyphosate, except that glyphosate kills the crops as well as the weeds. So genetic engineers found a gene which lets plants tolerate glyphosate, and transferred it into soybeans. Today, 63% of the soybeans grown in the US are glyphosate tolerant, allowing soil saving no-till agriculture on half the US soybean acreage.

There is another advantage to no-till agriculture. There are lots of plant residues beneath the ground, both root systems and humus transported by earthworms. Ploughing brings this material to the surface, where it can oxidize. Carbon dioxide is created, a greenhouse gas. So transgenic soybeans are a positive factor in postponing global warming. Approximately four tons of carbon dioxide are retained in the soil per acre per year. This saving is applicable to the accounting of CO_2 reductions in the Kyoto Treaty on Climate.

In fact, any technology which reduces the need to plough, spray, or till crops will reduce carbon dioxide emission. Consider a tractor pulling a ten foot wide harrow over a square mile of agricultural land. Simple arithmetic shows that the tractor will travel 528 miles, all the while burning gasoline.

Perhaps you are thinking that even if ploughing has disadvantages, herbicides don't sound very good either. The very word means "plant killer". But that is not the choice. Traditional soybeans are also grown using herbicides, most of which are far more toxic than glyphosate. On the average, the transgenic soybeans actually use 30% less total herbicide than conventional soybeans. So environmentally, this is a no-brainer.

Witchweed Control

Farmers in east Africa are plagued by a devastating parasitic weed called Striga, or witchweed.

Farmers are used to dealing with weeds that grow in the soil alongside the crop and compete for nutrients. From time immemorial, they have dealt with those weeds by pulling them up by hand. Less neighbour intensive methods like spraying and ploughing are now common. But none of these methods work for the witchweed. Striga attacks plants directly, underground, even before the weed has emerged above the soil surface. It sucks nutrients from the seeds and the roots of the crop. In some parts of Africa, the striga parasite destroys as much as 80% of the crop yield.

But now that a herbicide resistance trait can be transferred to a crop, scientists in Israel and Kenya, working together, have demonstrated a new strategy for striga control. Before planting the crop, they soak its seeds in a herbicide. The seeds are unharmed, but they become poisonous to the striga parasite. The seed germinates and sprouts without interference. By the time the crop is harvested, the herbicide has decomposed and disappeared.

Their demonstration used herbicide resistant transgenic corn. The same strategy would probably work with Africa's other important grains, sorghum and millet.

Soaking seeds would use far less herbicide than spraying it on the ground, and the complex spraying apparatus would not be needed. This is a significant consideration in Africa, where so many farmers are too poor to own expensive equipment.

Cheese Chymosin from Yeasts

Hard cheeses are made from whole milk by adding an enzyme called chymosin (rennet), which was formerly extracted from the stomachs of calves, a byproduct of veal. The gene for making chymosin was transferred from cows to yeast. Yeast can be grown in vats, as any brewer knows. Although many people consider it wrong to slaughter calves, yeasts have few defenders. Besides, chymosin from yeast is cheaper and purer than chymosin from calves. So today, almost all hard cheese (over 90%) is made from chymosin produced by genetic engineered yeast.

The poet Omar Khayyam wished for *a loaf of bread, a jug of wine, and thou beside me singing in the wilderness.* He owed two of his three pleasures to the working of yeasts. Today he would also be indebted to yeasts for a piece of cheese.

Cotton without Insecticides

Cotton farmers are plagued by various insect pests, such as the boll budworm, the tobacco budworm, and the pink bollworm. In the US south, where most of our cotton is raised, these insects were controlled using chemical insecticides. But there is a natural insecticide which has been used for almost a century by organic farmers, a bacterium called Bacillus thuringiensis, Bt for short. The bacterium produces a toxin which is deadly to caterpillars like the three mentioned above, but harmless to almost everything else (except insects of the order lepidoptera, butterflies and moths—even the legendary boll weevil (Anthromonus grandis) is not harmed by the Bt toxin). So genetic engineers transferred the gene for Bt toxin from Bacillus thuringiensis to cotton. Then the cotton plants, which could make Bt toxin, were cross bred with other varieties in the old fashioned way. Today, much of the US cotton crop is genetic engineered for the Bt toxin trait. The use of chemical insecticides in the cotton belt has declined dramatically; by over a million liters per year. Since the Bt toxin is inside the plant instead of sprayed onto the plant, the only insects which it can harm are those which eat the plant.

The benefit of reduced spraying of cotton is overwhelming. The cotton pesticides replaced are extremely damaging to the environment. They not only kill all insects in a cotton field, harmless or not, but also nearly anything else in the field, thus depriving insectivorous birds of their food. There is no way to keep these pesticides from getting into streams and rivers, where they are a serious hazard to aquatic life. We may think of cotton as a natural material, therefore environmentally friendly, but before there was Bt protected cotton, that was just wrong.

The gene for Bt toxin has been transferred into several other crops, including potatoes and corn. Approximately 30% of US corn is now transgenic, and the most popular transgenic varieties contain the Bt gene. Although we call it a toxin, to humans and other mammals and birds it's just a nutrient.

The principal potato pests are not caterpillars, but beetles, and the Bt toxin that protects cotton and corn doesn't harm beetles. But there is another

Bt toxin found in another variety of Bacillus thuringiensis which is deadly to beetles. The gene for that toxin was used in potatoes.

Biotechnology also has overzealous advocates who exploit consumers' fears about pesticides. Except in the case of serious accidents, there's little to worry about. Our bodies deal with many toxic substances in many foods, and as long as the amounts are small enough they give us no problems. The toxic load of pesticide residues on food is completely negligible in comparison with the toxins naturally present.

But as they are applied in the field, these agricultural pesticides are seriously hazardous. Each year many farm workers are poisoned by exposure to pesticides and farmers have nightmares about their children or pets being injured by playing near pesticides. Pesticides can also harm wild birds and small animals, and when they get into waterways they can kill fish and other aquatic life. The Bt transgenic plants reduce or eliminate this danger.

Slow Ripening Fruits

There are many fruits which ripen after picking. After they reach optimum ripeness, they begin to deteriorate. This is necessary for the life cycle of the plant, which relies on the sweet and pulpy parts to nourish the seeds. A ripe fruit literally digests itself. When this process is rapid, it effectively means that the fruit cannot be enjoyed out of season, or far from its growing area. For example, there is a popular Malaysian papaya variety which is unavailable outside Southeast Asia because it ripens so rapidly that it cannot be shipped very far. But it is quite easy to genetically engineer a fruit so that it does not ripen so rapidly. It doesn't even require a gene from another organism. Instead, a gene involved in the ripening process is copied with the message in reverse order. So now that plant has two genes with mirror image structure.

The way an organism uses the information in a gene to make a protein involves copying the gene (DNA) onto a messenger molecule, known as messenger RNA. The modified plant copies both the original gene and the mirror image gene to produce both types of messenger RNA. But since these messenger RNAs are exact complements of one another, they can wrap about one another just like the two strands of DNA, effectively blocking both messages. This means that the plant makes very little of the enzyme that causes ripening. This genetic engineering trick is called "antisense technology".

The Malaysian papaya was transformed in this way and therefore a slow ripening variety will soon be available.

The very first genetic engineered plant to be commercially developed as a whole food was a slow ripening tomato, called FlavR Savr. It was developed by Calgene, Inc. Because it could remain on store shelves for a long time, it could be left on the tomato plant until optimally ripe, and therefore the FlavR

Savr tomatoes sold for a premium compared to other tomatoes. Although consumers initially liked Calgene's tomatoes, they didn't ship well and the variety was eventually dropped.

Controlled Ripening

A coffee bush ripens a few coffee beans each day for many months. The best quality beans must be picked just after ripening, so picking coffee beans is very neighbour intensive. It would obviously be preferable if the beans would all get ripe at the same time.

Genetic engineering will make this possible. There is a coffee gene which turns on to initiate the last stage of ripening. Scientists modified a control gene so that the ripening gene does not turn on until the plant is sprayed with a triggering substance (patented and sold by the company that developed the coffee variety). Therefore all the beans on a bush reach the same not quite ripe stage and stop to wait for the triggering signal. The farmer decides when to spray the bush so it can be picked completely clean a few days later.

This can substantially improve the life of the small farmer. He can take a short vacation without losing part of his livelihood. He can work fewer hours per day, or he can pick all his crop in a few days and increase his income by working at another job. A large scale farmer would need fewer workers to pick the same quantity of coffee beans, and could afford to pay them a higher wage.

The control of when a crop is harvested would be valuable for other crops besides coffee. For example, the quality of grapes declines rapidly after they reach their optimum sugar content. Grape farmers now have to mobilize every available hand to harvest all their crop in a very short time. Their lives would be simpler if they could spread the harvest effort over a few weeks instead of a few days.

Large scale crops are harvested with special equipment. A farmer would not need to own a combine if he could rent it for the few days it was needed. But that wouldn't work if his neighbour needed to rent it for those same few days. If neighboring farmers could control when their crops become ready for harvest, they could share scarce and expensive equipment.

Saving the Banana

Wild bananas have seeds. They reproduce sexually, like beans and oak trees. But they aren't easy to eat. Bananas grown on plantations have no seeds. They are reproduced by taking cuttings from older banana plants.

Cultivated bananas are seedless because they have three of each type of chromosome instead of the normal two of each type. Such plants are called "triploid". They are always sterile. Genetic triploid freaks arise from time to time in nature, but modern breeders can also use chemicals or electric shocks to create triploid mutant cells.

Bananas have been cultivated for many thousands of years and there are about three hundred different banana varieties. Each variety was developed by crossbreeding wild bananas. Whenever a promising variety was been produced, the breeder caused it to be triploid, hence seedless. That plant was cloned by propagating cuttings, and it became the parent of its variety. All bananas plants of the same variety are genetically identical, like identical twins.

Of the three hundred varieties, only one single variety completely dominates international trade. It is called Cavendish. It is possible that you have never seen a banana other than a Cavendish banana.

Certain kinds of fungus can infect and kill banana plants. In many parts of the world, Cavendish banana plants are being attacked by a fungus called black Sigatoka. Since wild bananas can reproduce sexually, they are not all identical and some wild bananas can resist black Sigatoka. If the black Sigatoka fungus is present in a region, the resistant banana types become prevalent. But the Cavendish banana plants have no resistance. They are all identical so they all die.

To grow bananas commercially, growers must spray their plants with fungicides. Year after year, the black Sigatoka fungus has been evolving resistance to these fungicides, so growers have to spray more and more fungicide each year. Approximately one third of the cost of raising a banana is the cost of spraying it with fungicides, and it gets more and more costly each year.

A form of black Sigatoka banana disease, now spreading around the world, can tolerate all known fungicides. Soon it will attack bananas in Central America and the Carribean islands, the heartland of banana culture. The Cavendish banana will become virtually extinct. Agronomists estimate that this will happen within ten years.

This is not a fairy tale. It has happened before. Forty years ago, the most popular variety of banana was one called Gros Michel. But Gros Michel was susceptible to a fungus called "race 1 Panama disease". Now it is gone. Cavendish bananas, which are not susceptible to race 1 Panama disease, replaced them. (A related fungus, race 4 Panama disease, can infect Cavendish banana plants. At present it is only found in Malaysia and some nearby countries.) In turn, some other variety of banana could replace the Cavendish. It would look different and taste different but it would still be a banana.

There is only one practical way to save the Cavendish banana. It must be given some combination of genes from wild bananas which are not susceptible to the fungus. But this can't be done by any natural technique. Cross-breeding can create other banana varieties because wild bananas reproduce sexually. But Cavendish bananas reproduce only by cloning. Conventional breeding would have to rely on rare mutated Cavendish banana plants which can produce seeds and which can therefore be crossbred, in theory. Even the mutated plants produce only a tiny number of viable seeds, as few as two or three seeds in a hundred pounds of bananas. No banana breeding experts think that they can breed fungus resistance into a Cavendish banana variety in only ten years. Biotechnology can rescue the Cavendish banana. Scientists in Belgium have used genetic engineering to transfer some fungal resistance genes from wild bananas. The transformed Cavendish plants are not susceptible to the fungus and they can then be reproduced into nursery stock by the usual method of taking cuttings.

The Eggplant in Winter

The edible part of an eggplant is formed from the ovary of its flower. In this way, it is like the edible flesh of an apple, a pepper or grape. When we eat these fruits, we discard the seeds. But the plants only transform their ovaries into fruits when they start to produce seeds, although in the case of an eggplant, its seeds are so tiny that we ignore them. Eggplants will only set seeds in warm weather, so to grow them in the winter in an unheated greenhouse, the grower must use a chemical to trick the plant into beginning fruit development without setting seed. Such fruits do not grow very large or very fast under these conditions. So eggplants are expensive in the winter.

But now scientists in Italy have transferred two genes into a variety of eggplant, which not only allows the plant to set fruit in cool greenhouse conditions without chemicals, but also increases productivity of the same plant in either hot or cold weather.

The eggplant variety that the Italian scientists created is seedless. One of the two transferred genes is a switching gene which is turned on only in the ovary part of a flower. That gene turns on the other transferred gene, which makes a protein involved in synthesizing a growth hormone. The growth hormone makes the ovary grow into the fruit, just as it would have done in a traditional eggplant making seeds. Neither gene requires either seed setting or warm weather.

Where does one get seeds to produce large numbers of seedless eggplants? The transformed plants produce pollen, so they can be crossed with traditional eggplant varieties and the hybrid produced by that crossing has the seedless and self-starting property.

The scientists report productivity increases of 37% for the new variety, and they believe that the seedless type would be more marketable.

Virus Resistant Crops

Some viruses infect people or animals and other viruses infect plants. Plant viruses reduce the productivity of annual crops and can kill fruit trees.

Some plant viruses are spread by insects. Plants can be protected from those viruses by using insecticides or other pest management methods. There is essentially nothing else that a farmer can do to protect his crop from virus damage, except to grow a different crop. But genetic engineering a plant to protect it from a particular kind of virus is quite easy. A gene from the virus which encodes a protein in the virus' outer coat is copied into the plant's DNA. The plant then makes the coat protein, which is harmless, but which stimulates the plant's natural defences. Virus resistance traits have been introduced into many crops, including squashes, tomatoes, potatoes, tobacco and, perhaps most dramatically papaya.

Recently, cultivated Hawaiian papayas were hit by a devastating virus which essentially eradicated the commercial variety. Only the virus resistant transgenic papayas survived. If you like papaya, you can only buy the transgenic variety. Nobody can grow any other kind.

The Potato Famine

In 1840s Ireland, the potato crop was devastated by a late blight fungus (Phytophthora infestans) and Irish people starved en masse. That fungus could reappear at any time in any place and wipe out a potato crop.

Some varieties of potato have previously had some resistance to late blight fungus, but now a fungal strain has appeared in Russia that destroys those previously resistant varieties. This year a similar fungus appeared in potato fields in Prince Edward Island and 630 million pounds of potatoes, the island's principal crop, had to be destroyed.

But very recently, scientists were able to transfer a gene from alfalfa to a potato plant and the resulting potato plant is able to resist the fungus and thrive.

Potatoes also rot. A principal cause of potato rot is the bacterium Erwina carotovora, which has been called the flesh eating bacteria of the plant kingdom. Now a gene that confers resistance to E. carotovora has been coupled to a control gene that turns on when a plant has been wounded, and this construct has been transferred to experimental potatoes. As the researchers hoped, the modified potatoes, when punctured by a toothpick and exposed to E. carotovora, had almost twenty times less rot than unmodified potatoes.

Sentinel Crops

A recent innovation is a plant intended not for food but for quality control. It contains a gene derived from a luminescent jellyfish, but in all other ways it is identical to the food crop it is planted alongside. When these sentinel plants experience a lack of water, they literally glow in the dark. The farmer then knows that his crop must be watered or whether irrigation can be postponed.

In the western U.S. water is scarce. Agriculture is the biggest user of water.

Wasting water is intolerable. For example, so much water is taken from the Colorado River for irrigation that the river flows into Mexico a mere trickle, and it never gets to the sea at all. So this is yet another way that transgenic crops can benefit the environment.

Building with Silk

Silk is composed of two proteins, fibroin and sericin. The gene for fibroin has been transferred from a silkworm to a goat, and is expressed as a component of its milk. Soon we may also expect sericin to be transferred. It still remains to be seen whether technology can be developed to spin these proteins into a fibre.

That has already been accomplished with the kind of silk spiders use to make their webs. Genes for the two spider silk proteins were transferred to cells cultured from cow udders. Those cells then made the proteins. Happily, the spider silk can be spun by forcing a solution of its two protein components through a tiny nozzle.

The proteins self-assemble into spider silk strands. The same genes have since been transferred to live goats and when those goats are old enough to produce milk, it should be possible to make large quantities of spider silk very cheaply. Silk is an extraordinarily strong material, stronger than steel. In the future we may be getting our strongest building material from a farm instead of from a mine.

Safer Meat

Escherichia coli are friendly bacteria that live in our intestines and contribute to our health. But there is one strain of E. Coli (designated as O157:H7) that can make us sick, even kill us. We can get it from inadequately cooked meat. The E. coli infected meat comes from cattle with the virulent E. coli strain in their intestines. A cow's digestive system is adapted to digesting hay and grasses. The food first goes into a pre-stomach called a rumen. That's why cows are called ruminants. In the rumen, microorganisms turn the indigestible cellulose into nutrients the cow can assimilate. The food is then passed to a true stomach, and finally gets to the cow's intestines, where E. coli can live.

Therefore to guarantee against the virulent strain thriving in the cow's intestine, one needs to get some sort of prophylactic agent into the intestine. Antibiotics won't do. They would kill the cow's normal intestinal bacteria, and besides, it isn't a good idea to overuse antibiotics. There are antibodies specific to the virulent strain of E. coli, but they would be destroyed by passage through the cow's stomach before reaching its intestines. Genetic engineers are working on a neat solution. They are developing a transgenic animal feed which resists complete digestion in the stomach and delivers the antibody, specific to the virulent E. coli strain, into the intestines.

There are over 60,000 cases of E. coli illness in the United States each year. There would be many more except for an extensive programme of meat inspection. Even this understates the problem, because meat, if contaminated, has to be destroyed. E. coli infections were not such a serious problem when cattle were raised exclusively on grass and hay. On that diet, there isn't much digestion going on in the cow's intestine and the E. coli populations are comparatively low. But today's cattle spend the last weeks of their lives in feed lots, being fattened up on grain, which is digested in their intestines, leading to much higher populations of E. coli. So another way to solve the E. coli problem would be to raise leaner cattle and skip the feed lots.

Reduced Need for Fertilizers

One of the ways that farmers get better yields is by providing their plants with sources of organically bound nitrogen and phosphorus. These can be provided either by applying chemical nitrates or phosphates, or by using manures or decaying vegetation as sources of the same nutrients. Nitrogen is the largest constituent of the atmosphere, about 80%. It may seem paradoxical that unfertilized plants could suffer from a nitrogen deficiency while immersed in a sea of nitrogen gas, but it is just not available in the form they need. Of living things, only certain bacteria (and human chemists) have evolved the means to convert nitrogen from the atmosphere to a form useful to plants. But some plants, primarily legumes (peas and beans), have a symbiosis with these nitrogen fixing bacteria. The plants provide nodules on their roots that protect the nitrogen fixing bacteria, which then enrich the soil around those roots. Not only does this permit the legumes to grow luxuriantly without nitrate fertilization, but it makes the soil fertile for other plants growing in the same soil later. The technique of *crop rotation* is one of the oldest techniques of agriculture. Scientists hope to be able to transfer the genes which direct the formation of the nodules to other crops. If this is successful, the need for fertilizers would be dramatically reduced.

Unlike nitrogen, phosphorus is not a constituent of the atmosphere. There is no short-term likelihood that scientists will find a genetic engineering way to replace fertilizers that provide phosphates. The best hope for phosphate replacement would be to breed or engineer plants that make more efficient use of the phosphate available to them. If it proves impossible to engineer plants for nitrogen fixation, there are still options which can let them use fertilizers more efficiently. An enzyme called glucine dehydrogenase is involved in utilization of fertilizers.

The gene for glucine dehydrogenase is present in most crops, but it is expressed at low levels, because the control genes turn it off more than on. A genetic transformation of wheat which promoted increased synthesis of glucine dehydrogenase was 29% more effective in utilizing the same amount of fertilizer as the unmodified variety. The increased efficiency can either be used to grow more crop on the same land, or to cut down on the need for fertilizer to grow the same amount of crop.

More From The Sun

Plants derive energy from sunlight and use it to make sugar from carbon dioxide and water. This is photosynthesis. Scientists still do not have a complete understanding of how photosynthesis happens, although they know most of the steps. They know many of the genes which create the proteins needed for photosynthesis. They also know that there are differences in photosynthesis from one species to another. It happens that corn is an overachiever. Corn plants make more sugar per unit of sunlight than any of the other grains. An international team of scientists from Japan and from Washington State has transferred three of corn's photosynthesis genes into a rice plant. Early indications are that the transformed rice is more productive than the original rice variety.

A more important potential application may be the development of very fast growing trees. If global warming cannot be prevented by adding less carbon dioxide to the atmosphere, by burning less coal and oil, the only alternative is to depend on processes that remove it. Number one on that list is growing new trees.

Anything that makes agriculture more efficient can make more land available for growing trees. Anything that makes those trees grow faster removes more carbon dioxide from the atmosphere. For many environmentalists, preventing global warming is the highest priority. But proposed measures to restrict burning fossil fuels have encountered fierce political resistance. Opponents claim that such restrictions would be cost too much money, and would cost people their jobs, their comfort and their prosperity. By contrast, nobody loses anything if carbon dioxide is removed from the atmosphere by growing new trees.

Toxic Soils

Some soils are poor for plant growth because their mineral content is toxic. A high aluminum content is the most frequent problem, especially in acidic soils. But it has been possible to identify a few genes which enable some plants to extract aluminum compounds from soil and sequester them harmlessly in their fibrous parts. Recently, Florida scientists discovered a type of fern which can extract arsenic from the soil, although they do not yet know how the fern does this. But other teams have identified genes that can enable plants to remove cadmium, zinc and mercury from soils. By transferring such genes to fast growing plants, it should be possible to clean up some toxic soils in much the same way as we can use bacteria to clean up oil spills. In the nearer term, there is the work of Mexican scientist Luis Herrera Estrella. He transferred into corn a gene that allows the plant to overproduce a natural chemical, citric acid, which it then excretes through the roots. Citric acid binds to aluminum and prevents the plant from taking it up from the soil. Herrera's approach is not to extract aluminum from the soil but to prevent it from passing from the soil to the plant. A much larger problem is salt-contaminated soil caused by irrigation. Rainwater is very pure, but water borrowed from rivers contains some dissolved salt. Over many years of irrigation, the salt accumulates. But water cannot get from soil to roots if the soil water is saltier than the intracellular water. In fact, water goes the other way, from plant to soil, and the plant dies.

A gene was identified in a relative of cabbage. This gene enables the plant to pump salt from the soil into an isolated part of a cell, called a vacuole, where it is stored without harm to the plant. When salt is thus removed from the soil around the roots the plant can then take up the less salty water. The salt-tolerance gene was experimentally transferred to a tomato plant, where a control gene keeps it turned on all the time. The resulting tomato plant is able to grow well in salty soils. Happily, the fruit is not high in salt, but the plant's stems, leaves and roots are loaded with salt, so after the growing season the plant parts could be shipped elsewhere, making the soil become less salty each year. It's one more case of an environmental problem that can be solved by gene transfer.

Biological Pest Controls

Farmers, for very obvious reasons, would prefer not to use pesticides. They cost money and they are dangerous to use. Farmers much prefer *Integrated Pest Management* (IPM), a system that combines many different methods of suppressing crop pests, including encouraging predatory insects. Farmers even buy them. Agricultural distributors can supply such insects as ladybugs, praying mantids, lacewings and parasitic wasps. Integrated pest management includes using pesticides when other means are insufficient. But when a crop is sprayed with conventional insecticide, the harmful insects are not the only victims. Predatory insects may also be wiped out. Without any predators available, the pest populations can recover quickly, so that a second application of pesticide is required, which also kills the insect predators. This vicious cycle could be broken if genetical engineers can develop predatory insects resistant to common pesticides.

Rust Resistance

To a plant scientist, rust has nothing to do with oxidized metal. It is a plant disease caused by a fungus. It blights all the cereal crops, barley, wheat, oats, corn, millet and sorghum, *but not rice*. The rust fungus reproduces itself by forming club shaped cells called basidia. Each basidium bears four spores. When the spores are ready, they are released and carried by the wind.

The fungus infecting a single grain of wheat can easily produce millions of spores. The spores are so light that they can travel several times around the world before falling to the ground. Although some varieties are more resistant to rust than others, no variety is immune. But rice must contain some combination of genes that confers immunity to rust. If these genes can be identified and if their function can be deciphered, it should be possible to transfer them to other cereal crops and end, once and for all, this most important cause of famine.

Fast Growing Trees

Making paper requires large amounts of natural cellulose. Some can be derived from recycling, but most of our paper is made from freshly cut trees. The best trees for paper-making are fast growing softwoods with low resin content, like aspens. Genetic engineers have transferred genes for pest resistance and herbicide resistance into aspen and have tinkered with the genetic switches that promote growth to create a fast growing aspen that could supply our paper needs using considerably less land. The paper-making process must bleach out the brown colour of lignin, one of the components of wood. The bleaches used to be dumped in the nearest rivers, an important and highly visible kind of pollution. This is no longer allowed, but the disposal of chemicals from paper mills is still a major headache. At the Michigan Tech University, researchers have reduced the lignin content of aspen so that fewer chemicals are needed in the paper making process.

Fast Growing Fish

Most of the salmon we eat are caught wild, but some are grown in farm ponds. It takes about three years for a salmon to grow from fingerling size to optimum marketing size. In wild salmon a control gene turns on the gene for growth hormone, but only in the pituitary gland and primarily in warm water. So genetic engineers used a different control gene to turn on a growth hormone gene in cold water. That control gene was transferred from an ocean pout, and it originally turned on a gene for a protein that helped the pout tolerate very cold water.

The resulting creature looks and tastes just like the wild type salmon but it grows three times faster so it ought to be cheaper to produce. Wild salmon are now under environmental pressure from overfishing and because many of the streams where they lay their eggs are either polluted or inaccessible. If farmed salmon can economically replace more wild salmon, the pressure on this desirable species could be reduced dramatically.

There's a need for fast growing fish in rice growing regions. Rice is planted in standing water, but it is harvested from dry ground. With a slow growing rice farmers often raised fish in the rice paddy alongside the young plants. But newer strains of rice mature almost twice as fast as traditional varieties. Although this lets farmers grow more crops per year, unfortunately the rice paddies are not flooded long enough to raise fish. If genetic engineers were able to make fish grow faster, the farmers could again exploit this valuable protein resource.

Consumer Traits

Most of the traits in the examples mentioned so far have been targeted at the producer. The no-till soybeans are cheaper to grow because ploughing costs money. The cotton is cheaper to grow because chemicals cost money. The chymosin from yeast is cheaper than chymosin from calves. The consumer doesn't know how much water or fertilizer the farmer used, or whether a salmon is one year old or three. The vitamin enriched rice is the only one of the examples where the final product is better, rather than cheaper, for the person who eats it. But in the future we can expect to see "consumer traits". One of the first to appear will be potatoes genetically engineered to have a higher percentage of solids. If you love french fried potatoes but don't like the calories, you will love the new potatoes. They will absorb less oil but stay crispier longer.

Another valuable trait coming soon is coffee beans without caffeine. Caffeine is now removed from coffee by a chemical treatment, invented by German chemist Ludwig Roselius. Decaf is the only coffee I drink so I am looking forward to cheaper decaf coffee. In addition, many common foods are not safe for everyone. For example, peanuts cause a life threatening allergy in some people, especially children. Allergens are unusual proteins which are digested very slowly. Someone who has an allergy to a widely used food needs to read the small print on product labels, quiz the waiter in a restaurant, etc. If a child has the allergy, the problem is that much harder to manage. But peanuts or other crops are now being genetically engineered to eliminate these allergens. USDA scientists have identified the principal allergen in soybeans and have successfully developed modified soybeans which do not produce that allergen.

It is much easier to deactivate a gene, once its function is discovered, than it is to transfer a gene from one organism to another. The same antisense trick that was used to delay ripening can be used to suppress synthesis of an allergen. Therefore, once a gene has been identified which codes for an allergenic protein, the technology to eliminate that allergen from food crops is relatively easy, unless the allergenic protein is important to the life processes of the plant.

Lawns That Don't Need Frequent Mowing

In addition to food and fibre, genetic engineers are working to modify grass. Farmland constitutes mankind's biggest footprint on the earth, but lawns, athletic fields, golf courses, etc. have the biggest footprint in some communities. Environmentalists have many criticisms of lawns. They must be mowed regularly, which uses gasoline, creates noise pollution, and takes up people's time. In addition, the growing grass uses a very significant amount of water, fertilizers and pesticides, to make longer leaves which are then cut off by the mower. But genetic engineers are trying to develop a variety of grass which reaches a desired length and then dramatically slows its growth. Combined with pest resistance genes, this new grass would be almost as simple to maintain as astroturf.

What Can't be Genetically Engineered?

It takes only your imagination to come up with other possible applications of genetic engineering in agriculture. I like to joke about scientists developing a vegetable which grows its own bar code. But nothing can be accomplished until scientists have identified the relevant genes, figured out what they do, and figured out how the proteins they make work in the organism.

This point is vitally important. For example, scientists at one company tried to make a blue rose. They transferred the gene for delphinin, the blue pigment in delphiniums, into a rose and the rose made delphinin. But its flowers weren't blue. The scientists didn't understand how a delphinium uses its pigment, or how it would be used in a rose. There are virtually identical genes in flies and humans, but they have profoundly different results. Until scientists understand each step of some life process in one organism, they will never successfully transfer the trait to another organism. The only successes achieved so far have involved either a single gene or a group of closely related genes whose function has been worked out in detail. Some characteristics, such as humans' height, are affected by at least dozens of genes, some of which surely affect other attributes. We are very far from being able to genetically engineer complex traits. Unlike traditional plant and animal breeding, genetic engineering is not hit-or-miss. Genetic engineers do not have perfect control over the transferred genes, but they have much more control than the traditional breeders have.

TRANSGENIC PLANTS FOR ANTIBODIES

Immune response to produce antibodies against a pathogen is generally induced by an antigen, but as an alternative, antibodies can also be directly supplied from outside to provide for immunity. Therapeutic potential of antibodies, as a short-term relief against infectious agents, has been recognized for long, but their production has been difficult limiting their clinical use.

In 1989, for the first time, it was demonstrated that transgenic plants provide for an in expensive method for the production of functional monoclonal antibodies, also described some-times as plantibodies. Depending upon the crop, these plantibodies can be targeted to the seeds (cereals, oilsseeds, or legumes) or tubers (potato) which can be stored, transported and administered directly.

These recombinant antibodies include fully assembled whole immunoglobulins, antigen-binding fragments of immunoglobulins and single chain variable fragment (scFv).

Since several genes are involved in the synthesis of an antibody, transgenic plants each with a single gene are first produced and the genes are then brought together (gene stacking) by crossing such transgenic plants, each with a single gene. The resulting quadruple transgenics thus produced have been shown to assemble successfully the secretory immunoglobulin A (sIgA), which protects against microbial infection at mucosal sites.

For production of a single chain variable fragment in a. transgenic plant, gene constructs can also be prepared by joining together sequences encoding light and heavy chains of the immunoglobulin.

The production of transgenic plants expressing functional single-chain Fv protein (scFv), IgG and secretory IgA (sIgA) has already been achieved. These transgenic plants producing antibodies have applications in human d animal health care, since antibodies produced in plants have antigen binding ability, similar to the same protein expressed in the bacterial and mammalian systems.

The effectiveness of plantibodies against a variety of diseases has been demonstrated. For instance, plantibodies against cell surface antigens of

Streptococcus mutans has been shown to reduce tooth decay in animal and human models. Since large doses of an antibody are required for topical passive immunotherapy, transgenic plants may prove effective in producing these large quantities.

Transgenic Plants as Factories for Biopharmaceuticals

Cultured mammalian cells, bacteria .and fungi provided excellent production systems in the past for the production of biopharmaceuticals. However, with cultured mammalian cells are also sensitive to shear forces occurring during industrial-scale culture and also to the variation in temperature, pH, dissolved oxygen and certain metabolities.

This makes it necessary to control culture conditions thus adding to the cost. However, the bacterial and fungal systems are more robust, but they are not ideal for synthesis of many mammalian proteins, due to differences in metabolic pathways, protein processing, codon usage and the formation of inclusion bodies.

Transgenic plants have been considered as a better production system for the production of many pharmaceuticals to be used by humans and livestock, despite the fact that differences also occur between mammalian and plant systems with respect to codon usage and post translational processing. In view of this, many biotechnology companies are developing transgenic plants to be used as factories for the production of biopharmaceuticals.

For this purpose expression systems for high expression of introduced genes are being patented and transgenic plants produced using these expression systems are being already field tested. Clinical trials are also in progress using biopharma-ceuticals derived from these transgenic plants.

Like edible vaccines, the biopharmaceuticals derived from transgenic plants can be stored and distributed as seeds, tubers and fruits, which can be either directly used for oral ingestion or for extraction of the biopharma-ceutical.

This will make the immunization programmes in the developing countries cheaper and easy to administer, since the delivery by direct ingestion of modified plant product eliminates the need for product purification, which is an expensive process in pharma-ceutical industry.

Other advantages of using transgenic plants as production systems for the production of biopharmaceuticals include:

- Comparatively higher yields at a relatively low cost (according to some estimates, the production of a recombinant protein in provided excellent production systems in the past transgenic plants can be 10 to 50-fold cheaper than producing the same protein by E. Coli)

- Reduced health risk, which is common in the coventional pharmaceuticals
- Relatively little capital investment, due to the existing infrastructure.

The biopharmaceuticals that have been shown to express in transgenic plants include erythropoietin, insulin, enkephalins, a-interferon, human serum albumin, and two of the most expensive drugs, glucocerebrosidase and granulocyte macrophage colony-stimulating factor.

Many more therapeutic proteins will become available in future through investments currently being made in genomics research. The biotechnology company Applied a better production system for the production of Phytologics (API) in California (USA) produced transgenic rice producing a-I-antitrypsin, commonly used for treatment of cystic fibrosis liver disease, and hemorrhages.

Trials of a-I--antitrypsin extracted from malted grain of transgenic rice started in 1998 and it is hoped that regulatory approval for its commercial use should become available by 2004.

Production of Glucocerebrosidase for Gauchers Disease

One of the most expensive biopharmaceutical that has been shown to express in transgenic plants is glucocere-brosidase, which is an important lysosomal hydrolase enzyme, and whose deficiency causes Gauchers disease in humans. The drug for treatment of this disease is generally prepared from the enzyme purified from human placentas, and 10-12 tonnes of placentas per year are required for production of glucocerebrosidase needed for the treatment of a patient. This makes it the most expensive medicine in the world. In recent years, in order to reduce the cost, the drug has been produced in mammalian cell cultures, but it still remains one of the most expensive drugs. In the year 1999, the production of this drug in transgenic tobacco was patented by scientists at Virginia Tech and State University, and Crop Tech (both in Blacksburg, Virginia, USA). It is hoped that commercial production of this and other lysosomal enzymes in transgenic plants will be possible in future.

Production of Hirudin (an anticoagulant) for Thrombosis

Hirudin is another important drug that is used as anticoagulant for treatment of thrombosis. It was originally isolated from the leech (Hirudo medicinalis), but is now mostly produced by recombinant bacteria and yeast. Transgenic plants (oilseed rape, tobacco and Ethiopian mustard) were initially produced, which although expressed hirudin, but purification of hirudin from the seed proved rather expensive. It has been shown that hirudin in transgenic plants can be produced as a fusion protein with oleosin, which is transported to oil bodies thus making the purification cheaper and easier (oleosin-hirudin fusion proteins are extracted with oil bodies and can be easily separated by flotation centrifugation). Between the two genes (for hirudin and oleosin), the fusion gene also contained a sequence that encoded an endopeptidase recognition site, thus enabling easy cleavage of the fusion protein releasing hirudin from oleos in-hirudin fusion protein. This strategy also ensures that hirudin becomes active only after extraction, thus reducing environmental risks, if any. Transgenic oilseed rape expressing hirudin is now commercially grown by SemBiosys in Calgary (Canada).

TRANSGENIC PLANTS WITH BENEFICIAL TRAITS

During the last decades, a tremendous progress has been made in the development of transgenic plants using the various techniques of genetic engineering. The plants, in which a functional foreign gene has been incorporated by any biotechnological methods that generally are not present in the plant, are called transgenic plants.

As per estimates recorded in 2002, transgenic crops are cultivated world-wide on about 148 million acres (587 million hectares) land by about 5.5 million farmers. Transgenic plants have many beneficial traits like insect resistance, herbicide tolerance, delayed fruit ripening, improved oil quality, weed control etc.

Stress Tolerance

Biotechnology strategies are being developed to overcome problems caused due to biotic stresses (viral, bacterial infections, pests and weeds) and abiotic stresses (physical actors such as temperature, humidity, salinity etc).

Abiotic Stress Tolerance

The plants show their abiotic stress response reactions by the production of stress related osmolytes like sugars (*e.g.* trehalose and fructans), sugar alcohols (*e.g.* mannitol), amino acids (*e.g.* proline, glycine, betaine) and certain proteins (*e.g.* antifreeze proteins). Transgenic plants have been produced which over express the genes for one or more of the above mentioned compounds. Such plants show increased tolerance to environmental stresses. Resistance to abiotic stresses includes stress induced by herbicides, temperature (heat, chilling, freezing), drought, salinity, ozone and intense light. These environmental stresses result in the destruction, deterioration of crop plants which leads to low crop productivity. Several strategies have been used and developed to build ressitance in the plants against these stresses.

Herbicide Tolerance

Weeds are unwanted plants which decrease the crop yields and by competing with crop plants for light, water and nutrients. Several biotechnological strategies for weed control are being used *e.g.* the over-production of herbicide target enzyme (usually in the chloroplast) in the plant which makes the plant insensitive to the herbicide. This is done by the introduction of a modified gene that encodes for a resistant form of the enzyme targeted by the herbicide in weeds and crop plants. Roundup Ready crop plants tolerant to herbicide-Roundup, is already being used commercially.

The biological manipulations using genetic engineering to develop herbicide resistant plants are:

- Over-expression of the target protein by integrating multiple copies of the gene or by using a strong promoter.,
- Enhancing the plant detoxification system which helps in reducing the effect of herbicide.,
- Detoxifying the herbicide by using a foreign gene., and
- Modification of the target protein by mutation.

Some of the examples are:

Glyphosate resistance - Glyphosate is a glycine derivative and is a herbicide which is found to be effective against the 76 of the world's worst 78 weeds. It kills the plant by being the competitive inhibitor of the enzyme 5-enoyl-pyruvylshikimate 3- phosphate synthase (EPSPS) in the shikimic acid pathway. Due to it's structural similarity with the substrate phosphoenol pyruvate, glyphosate binds more tightly with EPSPS and thus blocks the shikimic acid pathway.

Certain strategies were used to provide glyphosate resistance to plants:

- It was found that EPSPS gene was overexpressed in Petunia due to gene amplification. EPSPS gene was isolated from Petunia and introduced in to the other plants. These plants could tolerate glyphosate at a dose of 2- 4 times higher than that required to kill wild type plants.
  - *By using mutant EPSPS genes*: A single base substitution from C to T resulted in the change of an amino acid from proline to serine in EPSPS. The modified enzyme cannot bind to glyphosate and thus provides resistance.
  - The detoxification of glyphosate by introducing the gene (isolated from soil organism- Ochrobactrum anthropi) encoding for glyphosate oxidase into crop plants. The enzyme glyphosate oxidase converts glyphosate to glyoxylate and aminomethylphosponic acid. The transgenic plants exhibited very good glyphosate ressitance in the field.

Another Example is of Phosphinothricin Resistance

Phosphinothricin is a broad spectrum herbicide and is effective against broad-leafed weeds. It acts as a competitive inhibitor of the enzyme glutamine synthase which results in the inhibition of the enzyme glutamine synthase and accumulation of ammonia and finally the death of the plant. The disturbace in the glutamine synthesis also inhibits the photosynthetic activity. The enzyme phosphinothricin acetyl transferase (which was first observed in Streptomyces sp in natural detoxifying mechanism against phosphinothricin) acetylates phosphinothricin, and thus inactivates the herbicide. The gene encoding for phosphinothricin acetyl transferase (bar gene) was introduced in transgenic maize and oil seed rape to provide resistance against phosphinothricin.

Other Abiotic Stresses

The abiotic stresses due to temperature, drought, and salinity are collectively also known as water deficit stresses. The plants produce osmolytes or osmoprotectants to overcome the osmotic stress. The attempts are on to use genetic engineering strategies to increase the production of osmoprotectants in the plants. The biosynthetic pathways for the production of many osmoprotectants have been established and genes coding the key enzymes have been isolated. *E.g.* Glycine betaine is a cellular osmolyte which is produced by the participation of a number of key enzymes like choline dehydrogenase, choline monooxygenase etc. The choline oxidase gene from Arthrobacter sp. was used to produce transgenic rice with high levels of glycine betaine giving tolerance against water deficit stress.

Scientists also developed cold-tolerant genes (around 20) in Arabidopsis when this plant was gradually exposed to slowly declining temperature. By introducing the coordinating gene (it encodes a protein which acts as transcription factor for regulating the expression of cold tolerant genes), expression of cold tolerant genes was triggered giving protection to the plants against the cold temperatures.

Insect Resistance

A variety of insects, mites and nematodes significantly reduce the yield and quality of the crop plants. The conventional method is to use synthetic pesticides, which also have severe effects on human health and environment. The transgenic technology uses an innovative and eco-friendly method to improve pest control management.About 40 genes obtained from microorganisms of higher plants and animals have been used to provide insect resistance in crop plants

The first genes available for genetic engineering of crop plants for pest resistance were Cry genes (popularly known as Bt genes) from a bacterium Bacillus thuringiensis. These are specific to particular group of insect pests, and are not harmful to other useful insects like butter flies and silk worms. Transgenic crops with Bt genes (*e.g.* cotton, rice, maize, potato, tomato, brinjal, cauliflower, cabbage, etc.) have been developed. This has proved to be an effective way of controlling the insect pests and has reduced the pesticide use. The most notable example is Bt cotton (which contains CrylAc gene) that is resistant to a notorious insect pest Bollworm (Helicoperpa armigera)..

There are certain other insect resistant genes from other microorganisms which have been used for this purpose. Isopentenyl transferase gene from Agrobacterium tumefaciens has been introduced into tobacco and tomato. The transenic plants with this transgene were found to reduce the leaf consumption by tobacco hornworm and decrease the survival of peach potato aphid.

Certain genes from higher plants were also found to result in the synthesis of products possessing insecticidal activity. One of the examples is the Cowpea trypsin inhibitor gene (CpTi) which was introduced into tobacco, potato, and oilseed rape for develping transgenic plants. Earlier it was observed that the wild species of cowpea plants growing in Africa were resistant to attack by a wide range of insects. It was observed that the insecticidal protein was a trypsin inhibitor that was capable of destroying insects belonging to the orders Lepidoptera, Orthaptera etc. Cowpea trypsin inhibitor (CpTi) has no effect on mammalian trypsin, hence it is non-toxic to mammals.

Virus Resistance

There are several strategies for engineering plants for viral resistance, and these utilizes the genes from virus itself (*e.g.* the viral coat protein gene). The virus-derived resistance has given promising results in a number of crop plants such as tobacco, tomato, potato, alfalfa, and papaya. The induction of virus resistance is done by employing virus-encoded genes-virus coat proteins, movement proteins, transmission proteins, satellite RNa, antisense RNAs, and ribozymes. The virus coat protein-mediated approach is the most successful one to provide virus resistance to plants. It was in 1986, transgenic tobacco plants expressing tobacco mosaic virus (TMV) coat protein gene were first developed. These plants exhibited high levels of resistance to TMV.

The transgenic plant providing coat protein-mediated resistance to virus are rice, potato, peanut, sugar beet, alfalfa etc. The viruses that have been used include alfalfa mosaic virus (AIMV), cucumber mosaic virus (CMV), potato virus X (PVX) , potato virus Y (PVY) etc.

Resistance against Fungal and Bacterial Infections

As a defence strategy against the invading pathogens (fungi and bacteria) the plants accumulate low molecular weight proteins which are collectively known as pathogenesis-related (PR) proteins.

Several transgenic crop plants with increased resistance to fungal pathogens are being raised with genes coding for the different compounds. One of the examples is the Glucanase enzyme that degrades the cell wall of many fungi. The most widely used glucanase is beta-1,4-glucanase. The gene encoding for beta-1,4 glucanase has been isolated from barley, introduced, and expressed in transgenic tobacco plants. This gene provided good protection against soil-borne fungal pathogen Rhizoctonia solani.

Lysozyme degrades chitin and peptidoglycan of cell wall, and in this way fungal infection can be reduced. Transgenic potato plants with lysozyme gene providing resistance to Eswinia carotovora have been developed.

Delayed Fruit Ripening

The gas hormone, ethylene regulates the ripening of fruits, therefore, ripening can be slowed down by blocking or reducing ethylene production.

This can be achieved by introducing ethylene forming gene(s) in a way that will suppress its own expression in the crop plant. Such fruits ripen very slowly (however, they can be ripen by ethylene application) and this helps in exporting the fruits to longer distances without spoilage due to longer-shelf life.

The most common example is the 'Flavr Savr' transgenic tomatoes, which were commercialized in U.S.A in 1994.

The main strategy used was the antisense RNA approach. In the normal tomato plant, the PG gene (for the enzyme polygalacturonase) encodes a normal mRNA that produces the enzyme polygalacturonase which is involved in the fruit ripening.

The complimentary DNA of PG encodes for antisense mRNA, which is complimentary to normal (sense) mRNA. The hybridization between the sense and antisnse mRNAs renders the sense mRNA ineffective.

Consequently, polygalacturonase is not produced causing delay in the fruit ripening. Similarly strategies have been developed to block the ethylene biosynthesis thereby reducing the fruit ripening. *E.g.* transgenic plants with antisense gene of ACC oxidase (an enzyme involved in the biosynthetic process of ethylene) have been developed. In these plants, production of ethylene was reduced by about 97% with a significant delay in the fruit ripening. The bacterial gene encoding ACC deaminase (an enzyme that acts on ACC and removes amino group) has been transferred and expressed in tomato plants which showed 90% inhibition in the ethylene biosynthesis.

Male Sterility

The plants may inherit male sterility either from the nucleus or cytoplasm. It is possible to introduce male sterility through genetic manipulations while the female plants maintain fertility. In tobacco plants, these are created by introducing a gene coding for an enzyme (barnase, which is a RNA hydrolyzing enzyme) that inhibits pollen formation. This gene is expressed specifically in the tapetal cells of anther using tapetal specific promoter TA29 to restrict its activity only to the cells involved in pollen production. The restoration of male fertility is done by introducing another gene barstar that suppresses the activity of barnase at the onset of the breeding season. By using this approach, transgenic plants of tobacco, cauliflower, cotton, tomato, corn, lettuce etc. with male sterility have been developed.

PRODUCTION OF POLYHYDROXYBUTYRATE OR PHB

Transfer of genes for acetoacetyl CoA reductase (phbB) and polyhydroxybutyrate (PHB) synthase (phbC), which catalyze two steps in the production of polyhydroxybutyrate or PHB (a biodegradable thermoplastic polymer) is another example 6f molecular farming. These genes were successfully transferred and their expression was demonstrated in transgenic plants of Arabidopsis thaliana.

Production of Human Serum Albumin (HSA)

Chimeric genes having CaMV35S promoter and encoding human serum albumin (HSA) were transferred successfully to produce transgenic potato and tobacco plants. The secretion of protein was achieved by using either the human preprosequence, or the signal sequence from extracellular PR-S protein from tobacco. HSA was secreted in transgenic leaf tissue

Production of Enkephalins

The production of pharmaceutically active compounds like enkephalins was also achieved in transgenic oil seed rape.

Production of Starch

Several species of higher plants are used for their stored starch. These include,

- Seeds of cereals and legumes and
- Tubers/roots of potato, yam and cassava.

In potato, 75% of dry weight is starch, making it a model system for improvement in starch quantity and quality. However, corn produces 17 million tons of starch as against potato producing 2 million tons per year; but transgenics in com are more difficult to produce and potato starch is certainly better.

There are two major components of starch, first, amylose (mol. wt. 104 to 106), which is a linear a(1-4) D-glucan polymer and second, amylopectin (mol. wt. 104), which is a branched a (1-4 and 1-6) D- glucan polymer. Besides these two components, starch also contains small amounts of lipids, proteins and phosphorus, which determine the starch quality. In higher plants, starch is found in two types of plastids, chloroplasts accumulating transitory starch (larger granules) and amyloplasts containing amylose. Amylose makes up to 11-37% of total reserve starch, but mutants with low and high amylose contents are known. The most important are the amylose free (waxy) mutants, which can be identified by staining with iodine (reddish brown colour in amylose free starch).

Starch is mainly used as a thickener in the food (*e.g.* sauce) and as sweetener in beverage industries (drinks, confectionery); about one third is also used either in paper, packaging and textile industry or as a raw material in chemical industry (production of ethanol, fructose and gluconate). For its industrial use, starch has to compete with cellulose or crude oil. It also needs to be modified according to need. For instance, starch from waxy maize resembles that from tapioca and has been used as its substitute, when tapioca was not available (Second World War). The demand for amylose free starch may increase, since it is easily digestible and can make clear pastes, that do not, retrograde (retrogradation means precipitation of amylose in aqueous solution of starch), when cooked as microwave-ready food. There is also a demand for high amylose crop plants for other purposes.

Starch is variously modified after isolation, 80% of potato starch being used only in a modified form (*e.g.* derivatized or gelatinized starches). Derivatized starch is used to improve paper strength and gelatinized starch is used to modify food texture. Thus starch production in crop plants needs to be modified for its variety of uses. Transgenic approaches may help in this direction.

Most of the starch in higher plants is synthesized from sucrose involving at least 13 enzymes, of which only the following three are considered as the key enzymes:

- ADP-glucose pyrophosphorylase (AGPase),
- soluble starch synthase (SSS)
- branching enzyme (BE).

A number of these genes in starch biosynthesis have been isolated and are available for use in the production of transgenic plants. Following results using techniques of genetic engineering have been obtained in potato.

- Antisense approach was successfully persued for inhibiting the action of granule-bound starch synthase (GBSS) enzyme. This led to 70%-100% inhibition of the activity of this enzyme resulting in decrease or complete absence of amylose content, thus giving amylose free starch.
- Starch content could be increased by introducing a bacterial gene

encoding AGPase (glgC) into potato, coupled with strong promoter belonging to patatin gene. This promoter induces high expression in tuber. DNA sequence for a transit peptide was also attached directing the starch to chloroplasts. Starch content was increased by at least 50%.

Production of Mannitol

Transfer of a gene for mannitol dehydrogenase from E. coli to tobacco was achieved, which led to increase in the level of mannitol in transgenic tobacco plants.

Transgenic Plants for Edible Vaccines

Genes for peptide epitopes of pathogens can be used for the production of vaccines against a variety of human and animal diseases. This concept of vaccine production in transgenic plants was introduced in 1992. In the year 1995, a biopharmaceutical company Prosigene from Texas (USA) conducted clinical trails, where pigs, fed with an edible maize vaccine, were found protected against the transmissible gastroenteritis virus (TGEV).

The company has also take patent on this and other edible vaccines produced in transgenic crops. Such vaccines, whenever successful, will provide a technology for production and delivery of inexpensive vaccines, especially in the developing world. This technology will also be cheaper than the expensive recombinant cell cuture-based expression systems.

A mucosal immune response is necessary particularly for combating infections due to bacteria/viruses invading epithelial membrane. Mucosal immune system involves production of secretory IgA (S-IgA) at mucosal surfaces such as gut and respiratory epithelia. Such a response is more easily achieved by oral vaccine rather than by parenteral (administered elsewhere than in alimentary canal) antigen delivery system. Particulate antigens rather than subunit or soluble antigens have been shown to be effective oral immunogens. The subunit vaccines whenever used require larger amounts (mg versus mg) of antigens and will be expensive if produced in recombinant cell culture expression system, which will need fermentation technology. In contrast to this, transgenic plants can be made to express the antigens in particulate form in the edible tissue, which can then be used as food, thus providing an inexpensive oral vaccine production and delivery system.

Factors Affecting Choice of Transgenic Plants for Vaccines

While using transgenic plants for vaccines, several factors need to be taken into consideration:

- It should stimulate a mucosal immune response and therefore diarrheal diseases were the first target for oral vaccine through transgenic plants;

- Antigen should assemble into ordered structures such as virus-like particles(vlp), which are resistant to digestion and reach the gut associated lymphoid tissue (galt), being perceived there are foreign antigen for production of antibodies;
- Plant system should be such that it may be used as food/feed, so that although tobacco was initially used, but it is not suitable. Potato and banana were subsequently used/ suggested, with an expression system, where antigen is produced or stored in the tuber in potato and in the fruit of banana. Alfalfa, coarse grain crops and beans can also be used for animal vaccines.

Transgenic Tobacco, Potato and Banana as Edible Vaccines

Several examples of successful production of prospective edible vaccines through transgenic plants are listed in the table. It may be seen that in majority of cases, transgenic tobacco plants were used, although other plants like potato, cowpea and blackeyed bean were also used.

In some of these cases, the oral immunogenicity was tested in mice and positive results were obtained and recovery of antibodies was possible.

It is hoped that eventually transgenic banana will be the fruit of choice of be used as edible vaccine, because banana is grown extensively throughout the world and can be consumed raw without any modification or processing of the edible part. The disease that will be targeted for production of edible banana vaccines may initially include measles, polio, diphtheria, yellow fever, viral diarrhea and hepatitis B, which will be successfully produced and used in the developing countries.

7

Plant Breeding

INTRODUCTION

The constant aim of plant breeding is to improve the quality, diversity and performance of agricultural and horticultural crops. The overriding objective is to develop plants better adapted to human needs. The evolution of plant breeding is a classic example of how improved biological understanding has been adapted to provide more effective methods of meeting the demands of a changing world. Plant breeding has existed in its most primitive form since the first farmers saved the seeds of their best plants from one season to the next more than 10,000 years ago. Over the centuries, this selection process has gradually become more scientific, bringing major improvements in the yield, quality and diversity of crops grown worldwide.

In the 19th century Gregor Mendel established the basic principles of plant genetics. Gregor Mendel in the 19th century laid the foundations for traditional plant breeding, in which selected parents are cross-pollinated to combine certain desired characteristics, such as high yield and resistance to pests and disease. He discovered that units of material, which are transferred from one generation to the next, determine inherited traits.

The plant breeder's aim is to reassemble these units of inheritance, known as genes, to produce crops with improved characteristics. In practice, this is a complex and time-consuming process. Each plant contains many thousands of genes, and the plant breeder is seeking to combine a range of desirable traits in one plant to produce a successful variety. Conventional breeding involves crossing selected parent plants, chosen because they have desirable characteristics such as high yield or disease resistance. The breeder's skill lies in selecting the best plants from the many and varied offspring. These are grown on and tested in subsequent years. Typically this involves examining thousands of individual plants for different characteristics ranging from agronomic performance to end-use quality.

Developing a new variety can take up to 15 years for wheat, 18 years for potatoes, even longer for some crops. The scope of conventional plant breeding has increased with improvements in technology. In the laboratory, chemical and mechanical techniques are used to speed up the selection process and remove natural barriers to cross-fertilization, for example between different crop species.

All forms of plant breeding involve the improvement of crops by selection. Since man began farming he has selected seeds from the best plants for the next generation. Over the centuries, this selection process has become more scientific, bringing major improvements in the yield, quality and diversity of agricultural and horticultural crops. Genes-units of hereditary material that are transferred from one generation to the next, determine these characteristics. Since the plant contains many thousands of genes, and the breeder is seeking to combine a range of traits in one plant, developing a successful variety can be an extremely lengthy process-up to 12 years in the case of cereals.

Using biological knowledge derived from cross-pollination, plant breeders have developed physical and chemically assisted ways of enhancing the speed, accuracy and scope of the selection process. Breeding systems involving grafting and hybridization, for example, have extended breeders' capabilities at the whole crop level, while more recent cell-based techniques enable breeders to operate at the level of individual cells and their chromosomes

A major objective in modern plant breeding is the making of crop varieties with the highest possible yield potential. Yield potential is defined as "the yield of a crop when growth is not limited by water or nutrients, pests, diseases, or weeds". For farmers, whose crops are indeed limited by such constraints, this yield concept may not be seen as the most relevant objective. Relieving crops of all sorts of environmental stress, leaves light and temperature, together with varietal characteristics, as the only determinants of yield. That achieved, the same high yielding variety can be used all over a climatic zone, and target area for the variety is enormous. If stress can not be eliminated, however, adaptation to a usually site-specific environment becomes necessary. In that case, the target area for each variety will be small.

Do these different perspectives warrant different approaches to plant breeding? Conventional wisdom says no. In experiments where varieties are tested at various levels of inputs, the same high yielding variety usually comes out as top yielder at all levels. Therefore, plant breeders often claim that their varieties not only perform excellently under high-input conditions, but would also be better than traditional varieties in a low-input environment. However, when these trials are taken outside of experimental farms and the varieties are tested under local or farm conditions, researchers discover what is called 'crossover in performance'. At a certain level of stress there is a crossover point beyond which local varieties or landraces perform better than the high yielding varieties Farmers who experience such situations are not a tiny minority.

To some degree, most if not all farmers have to cope with local stress conditions. The stress-free environment is hard to achieve, and even harder to sustain. In many favourable areas, farmers abandon the high-input technology for economic reasons or because of ecological problems, thus increasing the need for locally-adapted germplasm. But modern plant breeding, whether public or private, can not supply adapted germplasm everywhere. Only a system of local seed selection can ensure that. And that means devolution of plant breeding.

Can such decentralized breeding be compatible with development needs in a changing world and meet economic aspirations in a poor society? If the answer to these questions is going to be 'yes', the decentralized breeding must be able to take advantage of the power of science as well as of the capacities of local communities. Three aspects should be considered and made mutually compatible: breeding technology, participatory research methods, and organization at community level.

History

Plant breeding is the purposeful manipulation of plant species in order to create desired genotypes and phenotypes for specific purposes. This manipulation involves either controlled pollination, genetic engineering, or both, followed by artificial selection of progeny. *Plant breeding* often, but not always, leads to plant domestication.

Plant breeding has been practiced for thousands of years, since near the beginning of human civilization. It is now practiced worldwide by government institutions and commercial enterprises.

International development agencies believe that breeding new crops is important for ensuring food security and developing practices of sustainable agriculture through the development of crops suitable for their environment.

Domestication

Domestication of plants is an artificial selection process conducted by humans to produce plants that have fewer undesireable traits of wild plants, and which renders them dependent on artificial (usually enhanced) environments for their continued existence. The practice is estimated to date back 9,000-11,000 years. Many crops in present day cultivation are the result of domestication in ancient times, about 5,000 years ago in the Old World and 3,000 years ago in the New World. In the Neolithic period, domestication took a minimum of 1,000 years and a maximum of 7,000 years. Today, all of our principal food crops come from domesticated varieties. A cultivated crop species that has evolved from wild populations due to selective pressures from traditional farmers is called a landrace. Landraces, which can be the result of natural forces or domestication, are plants (or animals) that are ideally suited to a particular region or environment. An example are the landraces of rice, *Oryza sativa* subspecies *indica*, which was developed in South Asia, and *Oryza sativa* subspecies *japonica*, which was developed in China.

Classical plant breeding uses deliberate interbreeding (crossing) of closely or distantly related species to produce new crops with desirable properties. Plants are crossed to introduce traits/genes from one species into a new genetic background. For example, a mildew resistant pea may be crossed with a high-yielding but susceptible pea, the goal of the cross being to introduce mildew resistance without losing the high-yield characteristics. Progeny from the cross would then be crossed with the high-yielding parent to ensure that the progeny were most like the high-yielding parent, (backcrossing), the progeny from that cross would be tested for yield and mildew resistance and high-yielding resistant plants would be further developed. Plants may also be crossed with themselves to produce inbred varieties for breeding.

Classical breeding relies on homologous recombination of two genomes to generate genetic diversity. The Classical plant breeder may also makes use of a number of *in vitro* techniques such as protoplas fusion, embryo rescue or mutagenisis to generate diversity and produce plants that would not exist in nature.

Before World War II

Intraspecific hybridization within a plant species was demonstrated by Charles Darwin and Gregor Mendel, and was further developed by geneticists and plant breeders. In the early 20th century, plant breeders realised that Mendel's findings on the non-random nature of inheritance could be applied to seedling populations produced through deliberate pollinations to predict the frequencies of different types.

In 1908, George Harrison Shull described heterosis, also known as hybrid vigour. Heterosis describes the tendency of the progeny of a specific cross to outperform both parents. The detection of the usefulness of heterosis for plant breeding has lead to the development of inbred lines that reveal a heterotic yield advantage when they are crossed. Maize was the first species where heterosis was widely used to produce hybrids.

By the 1920s, statistical methods were developed to analyse gene action and distinguish heritable variation from variation caused by environment. In 1933, another important breeding technique, cytoplasmic male sterility (CMS), developed in maize, was described by Marcus Morton Rhoades. CMS is a maternally inherited trait that makes the plant produce sterile pollen, enabling the production of hybrids and removing the need for detasseling maize plants.

These early breeding techniques resulted in large yield increase in the United States in the early 20th century. Similar yield increases were not produced elsewhere until after World War II, the Green Revolution increased crop production in the developing world in the 1960s.

After World War II

Following World War II a number of techniques were developed that allowed plant breeders to hybridize distantly related species, and artificially induce genetic diversity.

When distantly related species are crossed, plant breeders make use of a number of plant tissue culture techniques to produce progeny from other wise fruitless mating. Interspecific and intergeneric hybrids are produced from a cross of related species or genera that do not normally sexually reproduce with each other. These crosses are referred to as *Wide crosses*. The cereal triticale is a wheat and rye hybrid. The first generation created from the cross was sterile, so the cell division inhibitor colchicine was used to double the number of chromosomes in the cell. Cells with an uneven number of chromosomes are sterile.

Failure to produce a hybrid may be due to pre- or post-fertilization incompatibility. If fertilization is possible between two species or genera, the hybrid embryo may abort before maturation. If this does occur the embryo resulting from an interspecific or intergeneric cross can sometimes be rescued and cultured to produce a whole plant. Such a method is referred to as *Embryo Rescue*. This technique has been used to produce new rice for Africa, an interspecific cross of Asian rice *(Oryza sativa)* and African rice *(Oryza glaberrima)*. Hybrids may also be produced by a technique called protoplast fusion. In this case protoplasts are fused, usually in an electric field. Viable recombinants can be regenerated in culture.

Chemical mutagens like EMS and DMSO, radiation and transposons are used to generate mutants with desirable traits to be bred with other cultivars. Classical plant breeders also generate genetic diversity within a species by exploiting a process called somaclonal variation, which occurs in plants produced from tissue culture, particularly plants derived from callus. Induced polyploidy, and the addition or removal of chromosomes using a technique called chromosome engineering may also be used.

When a desirable trait has been bred into a species, a number of crosses to the favoured parent are made to make the new plant as similar as the parent as possible. Returning to the example of the mildew resistant pea being crossed with a high-yielding but susceptible pea, to make the mildew resistant progeny of the cross most like the high-yielding parent, the progeny will be crossed back to that parent for several generations. This process removes most of the genetic contribution of the mildew resistant parent. Classical breeding is therefore a cyclical process.

It should be noted that with classical breeding techniques, the breeder does not know exactly what genes have been introduced to the new cultivars. Some scientists therefore argue that plants produced by classical breeding methods should undergo the same safety testing regime as genetically modified plants. There have been instances where plants bred using classical techniques have been unsuitable for human consumption, for example the poison solanine was accidentally re-introduced into varieties of potato though plant breeding.

Exploiting heterogeneity and crop evolution in farmers' fields are outside the scope of most plant breeding research. One exceptional experiment, however, has shed some scientific light on the issue. It was started at the University of California (UC) in 1928. Composite cross populations of barley were produced, some of which were extremely diverse in origin of sources. These populations were exposed to continuous natural selection in current modern farming environments and became the subject of studies during the career span of several generations of UC professors.

It appears that after low yields in initial years, the composite cross populations gradually improved in performance and eventually became quite good yielders, with excellent yield stability and disease resistance. These results inspired Suneson to propose an evolutionary plant breeding method. After assessment of later generations of the same material, Soliman and Allard concluded that such evolutionary breeding "is unwarranted" if yield potential is the major goal. However, if disease resistance and yield stability are two major objectives, "the composite cross approach is an efficient method". This amounts to saying that a major part of world agriculture, many high-input systems included, could be well served by this approach.

In short, this means constructing a body of broadly diversified germplasm and exposing it to natural selection in areas of contemplated use. For those who are familiar with traditional farmers' breeding, this may sound like reinventing the wheel. In fact it is an improvement of the old wheel of plant breeding.

The first step, constructing a body of of broadly diversified germplasm, is not all that straightforward. Science has access to world collections and information sources that are unavailable to farmers. A research institute can chose relevant germplasm and make composite cross populations with an evolutionary potential, most probably far beyond that of locally available varieties.

The immediate outcome, the early generation composite cross population, will be unadapted everywhere and is likely to yield poorly. With time, however, recombinations and natural sorting will improve the adaptation, and, according to the Californian experience, narrow the gap with commercial varieties. The long term outcome could be populations that outperform commercial varieties in disease resistance and yield stability and that may be used as a source of artificial selection for high yield.

The disease resistance appears to have evolved through the building up of polygenic complexes. Therefore, it provides a durable resistance as opposed to the monogenic and, therefore, mostly non-durable resistance usually bred into commercial pure line varieties. The stability, at least to some degree, depends on the buffering effect of crop heterogeneity. The Californian experiment shows that when the population is propagated in isolation for a very long time, diversity will start declining, resulting eventually in reduced stability.

These experimental findings into the context of current development needs, a few conclusions can be drawn. We need breeding populations with a very high evolutionary potential, and these populations must be exposed to the stress conditions of, or similar to, current farm environments. Furthermore, a certain level of diversity within populations must be maintained in order to sustain evolutionary potential and yield stability.

An Age-old Tradition

Modern plant breeding is a sophisticated, high investment business, but its origins stretch back thousands of years to primitive farmers who selected the best plants in one year to provide seed for their next crop. This selective breeding was the first human refinement of natural plant evolution. Recent scientific and technological developments have allowed a greater rate of improvement. It was Gregor Mendel who, in the mid- 19th century, first provided a scientific explanation of genetic inheritance. His conclusions on the relationship between inherited characteristics in the offspring and the genetic makeup of the parents were the theoretical basis for classical plant breeding. Mendel's work went largely unrecognized in his own lifetime, and it was not until the early 20th century that it was rediscovered to become the basis of modern scientific plant breeding.

A Modern Industry

Until the early 1960s, plant breeding in Britain was largely confined to publicly funded research. This situation changed dramatically in the mid-1960s, with the passing into UK law of the 1964 Plant Varieties and Seeds Act. This legislation introduced a system of royalty payments on individual plant varieties, known as Plant Breeders' Rights, and triggered a rapid expansion of plant breeding as a commercial enterprise in its own right. Today, much of the basic research into crop science is still conducted by public sector research organisations, but the majority of commercial plant breeding takes place within the private sector. Some 60 plant breeding companies, based in the UK, are active across the entire spectrum of plant species from the major arable crops through to ornamental garden shrubs and flowers. In total, the plant breeding sector employs around 5,000 people, and supports a further 5,000 jobs in seed production and distribution. Plant breeding remains a vital industry to keep Britain competitive on world markets. The need for new varieties, adapted to our unique growing conditions, is never ending, driven by the challenges of new disease pressures, changing market requirements and shifts in agricultural and environmental policies.

- *Maize*: Maize crops used for grain and animal fodder are derived from wild races originating in Central America
- *Sugar beet*: Modern varieties of sugar beet have been developed from wild ancestors native to Central Europe.
- *Potatoes*: Wild ancestors of the modern potato still grow in parts of South America.
- *Wheat and Barley*: Small grain cereals, the mainstay of UK crop production, are derived from wild grasses of the Middle East.
- *Oilseed Rape*: Like many crops within the brassica family, oilseed rape has its origins in wild species native to China.

PLANT BREEDING FOR A WORLD COMMUNITY

There is enormous potential for plant breeding to benefit less developed parts of the world, where food security is critical as populations continue to increase.

In particular, advances in genetic technology may offer more versatile solutions than conventional systems of crop improvement. For example, scientists in Britain are pioneering important research to develop salt and drought tolerance in crop plants. This may help to alleviate the devastating effects of crop failure in arid regions such as sub-Saharan Africa. Private and public sector initiatives to share knowledge and expertise are already in place. They include commitments to provide Vitamin A enhanced rice and virus resistance in sweet potatoes free of charge to developing countries.

International arrangements to recognize genetic resources, as the sovereign property of individual nation states will stimulate further programmes of technology transfer and benefit sharing between industrialized and less developed regions of the world.

Effects of Modern Plant Breeding on Environment

Some environmentalists are concerned that genes from genetically modified crops could escape and transfer to other species with unwanted consequences. For example, it is argued that herbicide-resistant crops could cross with weedy relatives to create a new strain of 'super weed'. In practice, since the reproductive systems of genetically modified and conventionally bred crops are identical, the behaviour of domesticated crop plants is unlikely to be affected by single gene changes.

In ten years of worldwide field trials there have been no adverse reports of genetically modified crops spreading in the environment. Furthermore, resistance to weed killers has been available for decades in a number of conventionally bred varieties without causing any such problem. The development conventionally bred crops-nevertheless the technology is subject to much tighter controls. In the long history of plant breeding, the strict regulations applied to the development and use of genetically modified crops is unprecedented. In both the laboratory and the field, each genetically modified crop must go through a rigorous process of monitoring and evaluation before it can reach the customer. Extensive and ongoing evaluation of genetically modified crops has shown that the technology presents no new food safety risks. Indeed, biotechnology will enable breeders to develop food crops with improved nutritional value and better keeping qualities. Enhanced disease and pest resistance will also reduce pesticide residues.

SPEEDING-UP PLANT-BREEDING WORK

Plant-breeding work involving crosses generally takes a great number of generations before the superior new plant has been obtained. In the routine work of a large plant-breeding station so many breeding projects arc under way simultaneously that it matters little how much time elapses between the initial cross and the final purification of a promising commercial variety. Every year there are many projects being started, several are under way, and a few arc in their last stages.

It often pays to cut down the time required for work of this kind. One of the most promising means of doing this is to grow more than one generation of plants in one year. This usually means that one of those generations of plants is grown in a season in which the plant is not ordinarily grown, and that for this reason a correct evaluation of the quality of the individual plants is out of the question.

In Europe or in America growing an extra generation on the spot means growing plants in the greenhouse. To grow cereals or beans in winter in greenhouse conditions not only requires heat; we must also have the means of regulating the length of day. The simplest method is to use large electric lights to furnish both the heat and the light, to lengthen the day by those means, and to let the temperature go down during the dark part of the night. Under those conditions cereals and beans will flower and set seed very well in the winter months,

The electricity required per square yard is roughly that used by lamps that require a kilowatt. It is clear that under such conditions the winter generation will not show the normal qualities of the plants in that group which are grown in summer-time in the field, but there are many cases in which this is not of much importance.

If we want to proceed by the method of making some hybrids and then growing a second and third inbred generation from this stock, we can do our crossing in the field, grow our hybrids in the winter, and then grow another crop of second-generation plants in normal field conditions. When we are using the method of mating back hybrid stock to some pure kind of plants we want to improve it matters but little if we cannot very well select the best plants during some of the generations involved. It is evident that we must grow a generation under normal field conditions whenever we want to make our selections.

In recent years a few seed-firms have hit upon the perfect scheme of growing two generations of annual plants. This does not involve any greenhouse generations. They work in two different localities with very similar climate, but so chosen that one of those localities is situated in the Southern Hemisphere.

There are regions of Chili where Californian firms can find conditions of plant-growth that are identical with conditions at home, with the exception that the growing season comes during the Californian winter. In those circumstances it is quite possible to grow two absolutely normal generations during one year. Our present greatly improved methods of communication make such schemes perfectly feasible.

With many tropical plants one generation yearly is grown of plants that need special conditions of either a wet or a dry season. Here it is easy enough to interpolate an additional generation, by looking for a suitable spot, or by providing the necessary irrigation. Sometimes it takes a very long time to bring the plants to a condition where we can evaluate them. A good example is that of the fruit trees. It takes many years for an apple seedling or a seed-grown cherry to come into bearing. Here we can save a great deal of time by budding or top-grafting parts of the seedlings upon mature trees in full bearing. By this method we may save more than half the number of years normally required for bringing the seedlings to maturity.

In this connection we might also treat of those cases where certain seeds take a very long time to germinate. Storing the seed in suitable conditions of moisture, and especially of temperature, will speed-up germination in such plants as gooseberries and celery. In stone-fruits it has been found possible to crack open the stone, to extract the kernel and grow the seedlings under aseptic or even under antiseptic conditions (thymol solution).

Shows and Showing

The exhibiting of plants and seed at the shows is a very curious thing. It pays extremely well for a seed-firm to employ some gardeners who know just exactly when and' how to get enough plants full of luscious beans or covered with flowers on the exact day for each show. Florists and lovers of gardening see the quality of such plants at the shows, and many people who admire them there try in their turn to win the coveted prizes at their local shows the next year. This show game is certainly very nice and pleasant; it gives the gardeners something to strive after; it may even help to keep some of the lads in the country.

On the other hand, there is very, very little connection between the qualities judged at the shows and actual value of the material shown. The few selected mangolds, or the mammoth pumpkins, or the bundle of wheat- ears, or even of the maize-ear contests of the American agricultural shows.

All along the line, in horticulture as well as in agriculture, the selected samples shown give no indication of the value of the seed or plant stock the producers have for sale. Of course we can see that those few mangolds have the ideal shape and size, but the seed-firm must give us some guarantee about the percentage of mangolds that will grow to this size and shape from the seed they sell, just as the owner of a prize-winning bull must give some guarantee of this animal's ability to give profitable daughters, in addition to the blue ribbon awarded to the beautifully coloured and shaped prize-winner.

In agriculture this is beginning to be so well realised that the classes for beautiful corn cobs or for the largest mangolds are becoming very rare indeed, even if we still see stock-judging in dairy cows and similar farces. In the horticultural shows much interest is still shown in regard to flowers and vegetables.

It seems evident that when we strive for beauty, beauty contests for roses and begonias, for asparagus and delphiniums and dahlias are indicated. The show is a shop-window. The seller must show if his competitors show, if only to keep his name before the public. To the public the show is only just a pleasant occasion for a half-day outing. To the buyer the show is much nicer to see than a description in a catalogue.

Long lists of novelties are bought at the shows by people with an experimental turn of mind. We see the representatives at the shows kept busy noting down long lists of orders. Showing certainly pays the seller. A very much more sensible system of showing a great many varieties is that of the demonstration gardens and experimental stations, both stations in which the agricultural authorities try to compare promising new and good old varieties with the aid of objective standards, and such gardens as show-plots arranged by rose-lovers or by societies of growers of gladioli or dahlias. Here the varieties can be judged throughout the growing season in favourable conditions.

The Annual Self-fertilizing Crops

The number of plant species now being cultivated by man is really enormous, while only relatively few animal species are kept under domestication. It is possible for the author of a book on animal breeding to take those species almost one by one, and to give the specialist breeders a few hints over and above those which are contained in the general book. This would be quite out of the question in a book on plant breeding. One way out of the difficulty would be to give one example of each of the three groups—self-fertilized plants, cross-fertilizers and plants that are usually propagated by vegetative means. These plants in two sections— grains and legumes—giving some hints to the breeders of wheat, barley, oats, sorghum, rice, etc., and going into some details about the breeding of beans, soybeans, cow-peas, ground-nuts, peas, etc., separately. But, after all, the self-fertilizing annuals have so much in common from the point of view of plant-breeding methods.

Self-fertilization causes isolation. Each plant, as far as reproduction is concerned, is as self-contained and isolated in a crowded field as it would be if grown by itself in a greenhouse in a city. The inbreeding caused by self-fertilization has several results. It makes all heritable variability disappear in every line, it makes every group consist of parallel lines, in which each plant descends from only one parent plant, so that the usual complicated network of ancestry is simplified into a system of isolated parallel lines. Where inherited variability exists, such as happens after a deliberate artificial cross or when accidental cross-fertilization occurs, it will automatically disappear. That is to say, it will disappear in so far as each separate line is concerned, for Mendelian segregation will make the lines, descending from any hybrid heterozygous for a great many genes, diverse in their hereditary make-up. The result is that a field of self-fertilizing annuals that has not been grown from just one pure line by a plant breeder always consists of a mixture of a number of genetically different lines. Even if accidental cross-fertilization or irregularities in crossing-over (mutation) are very rare, they do occur, and cause this state of things. This genetic variability, however, is wholly different from what we find in other plants and animals. We are dealing with mixtures of pure lines, and almost every plant we take from any field is pure, homozygous for all its genes, and will, if its seeds are sown, give a very homogeneous descendance. From all this it follows that enormous progress can often be made in all self-fertilized plants by simply picking out a great number of promising-looking individual plants and comparing their descendance.

Many valuable widely grown wheats, tobaccos, peanuts and beans have been just found. They "just growed ", like Topsy. The first step in plant-breeding in the self-fertilizing plants must always be to look for those ready-made pure lines. The next is to find whether a mixture of two of the very best pure lines can be discovered which will give us still higher production (10 per cent, higher production from a mixture of two strains as compared with that of the best one is quite common).

We have then to produce some more genetic variability by deliberate cross-breeding. This generally means that we try to combine the good qualities of two different strains in one new one. Here there are two wholly different schools of thought, and two wholly different methods. One of them is to analyse second-generation plants, and to find those individuals that do combine, let us say, the disease resistance of line A with the large quantity of big seeds of line B. This is a method in- vented by the experimental geneticists who are trying to combine practical plant breeding with purely scientific gene- analysis. As a theoretical geneticist, interested in genes and their action. It is great fun playing with a hundred different hybrid lots in one species, growing a few dozen second-generation plants in each lot and tabulating the results. But this is neither sound genetics nor sound plant breeding. We are dealing with so many unanalysed and unanalysable genes which all affect production and quality at the same time, that the chance of finding something really worth while in a few hundred plants is simply not good enough.

The other method is based upon the realization that hundreds of genes are commonly involved, and upon the fact that purification—reduction of the potential variability—is rapid and automatic in this material. It consists of growing the greatest possible number of seeds from first-generation hybrid plants. We can then proceed in two different ways. One is an analytical method, which means that we are seeking for the very best-looking plants (for instance, by weighing their seeds), and progeny testing those plants by sowing a large number of seeds from each, in order to find the most profitable families. The other method is what Baur and Nilsson-Ehle called a "Ramsch" method.

This consists simply of growing the descendance of a hybrid in a vast mixture, refraining from any analysis, and continuing this hybrid lot as a miscellaneous collection for five or six generations, sowing as much of the mixed seed of each harvest as we have room for. If we do this, natural selection will weed out most of the unproductive and undesirable types, such as those plants t1 at ripen after harvesting time, all weaklings and all plants susceptible to diseases and pests.

After five or six years that lot will consist of a mixture of pure lines, every line different from every other, with the successful lines represented by more individuals than the unsuccessful ones. After this we will treat the mixture just as if we were dealing with any other mixture, growing separate rows of beds each from one likely-looking plant.

This is a very sound method, which saves a great deal of un- necessary labour and time. It is extremely difficult to find just how much better this method is as compared with that of analysis in every generation, starting from the second generation. An analytical method could be made to work very well, provided we did all we could to make the number of plants of the second and third generations exceedingly large (in the cereals we can get almost as many seeds as in tobacco by splitting up the hybrid plant repeatedly before it starts to grow any stalks).

In actual practice there are only a few crosses that really produce anything worth while, and the general practice seems to be to make a great many attempts and grow a large number of first-generation hybrids.

This renders it imperative to cut down on the number of plants per experiment. It is a vicious circle. This is all very wrong and wasteful. One cross every year and work with it, both analytically and by means of a "Ramsch ", doing the final selection after six years for one lot in every season, than waste my time castrating flowers and producing numerous hybrids that would have no really good chance of showing what they could do.

In all the self-fertilizing annuals we must be continually on the look-out for qualities that may be of special merit. Mostly of adaptations to special conditions.

In wheaa on irrigated ground, resistance to drowning may be as important as disease resistance or quality of the grain. Conversely, rice could be grown much more extensively if we would take the trouble to do some plant-breeding work on dry-land rices, which abound in the mountain villages of all the tropical Asian islands.

In ground-nuts we should look for lines that are specially adapted to mechanical harvesting, as well as for lines adapted to the production of a first-quality product in the hands of tens of thousands of small cultivators.

In all the plants of this group it is possible not only to improve the yield and the quality, but also to extend the area in which profitable production is possible. Tobacco, rice, soybeans and cow-peas are just as important in this respect as wheat, barley and oats.

BREEDING METHODS IN CROP PLANTS

Mass Selection

In mass selection, seeds are collected from (usually a few dozen to a few hundred) desirable appearing individuals in a population, and the next generation is sown from the stock of mixed seed.

This procedure, sometimes referred to as phenotypic selection, is based on how each individual looks. Mass selection has been used widely to improve old "land" varieties, varieties that have been passed down from one generation of farmers to the next over long periods.

An alternative approach that has no doubt been practiced for thousands of years is simply to eliminate undesirable types by destroying them in the field. The results are similar whether superior plants are saved or inferior plants are eliminated: seeds of the better plants become the planting stock for the next season.

A modern refinement of mass selection is to harvest the best plants separately and to grow and compare their progenies. The poorer progenies are destroyed and the seeds of the remainder are harvested.

It should be noted that selection is now based not solely on the appearance of the parent plants but also on the appearance and performance of their progeny. Progeny selection is usually more effective than phenotypic selection when dealing with quantitative characters of low heritability. It should be noted, however, that progeny testing requires an extra generation; hence gain per cycle of selection must be double that of simple phenotypic selection to achieve the same rate of gain per unit time.

Mass selection, with or without progeny test, is perhaps the simplest and least expensive of plant–breeding procedures. It finds wide use in the breeding of certain forage species, which are not important enough economically to justify more detailed attention.

Pure–Line Selection

Pure–line selection generally involves three more or less distinct steps:

1. Numerous superior appearing plants are selected from a genetically variable population;
2. Progenies of the individual plant selections are grown and evaluated by simple observation, frequently over a period of several years;
3. When selection can no longer be made on the basis of observation alone, extensive trials are undertaken, involving careful measurements to determine whether the remaining selections are superior in yielding ability and other aspects of performance.

Any progeny superior to an existing variety is then released as a new "pure–line" variety. Much of the success of this method during the early 1900s depended on the existence of genetically variable land varieties that were waiting to be exploited. They provided a rich source of superior pure–line varieties, some of which are still represented among commercial varieties. In recent years the pure–line method as outlined above has decreased in importance in the breeding of major cultivated species; however, the method is still widely used with the less important species that have not yet been heavily selected.

A variation of the pure–line selection method that dates back centuries is the selection of single–chance variants, mutations or "sports" in the original variety. A very large number of varieties that differ from the original strain in characteristics such as colour, lack of thorns or barbs, dwarfness, and disease resistance have originated in this fashion.

Hybridization

During the 20th century planned hybridization between carefully selected parents has become dominant in the breeding of self–pollinated species. The object of hybridization is to combine desirable genes found in two or more different varieties and to produce pure–breeding progeny superior in many respects to the parental types. Genes, however, are always in the company of other genes in a collection called a genotype. The plant breeder's problem is largely one of efficiently managing the enormous numbers of genotypes that occur in the generations following hybridization.

As an example of the power of hybridization in creating variability, a cross between hypothetical wheat varieties differing by only 21 genes is capable of producing more than 10,000,000,000 different genotypes in the second generation. At spacing normally used by farmers, more than 50,000,000 acres would be required to grow a population large enough to permit every genotype to occur in its expected frequency. While the great majority of these second generation genotypes are hybrid (heterozygous) for one or more traits, it is statistically possible that 2,097,152 different pure–breeding (homozygous) genotypes can occur, each potentially a new pure–line variety. These numbers illustrate the importance of efficient techniques in managing hybrid populations, for which purpose the pedigree procedure is most widely used.

Pedigree Breeding

Pedigree breeding starts with the crossing of two genotypes, each of which have one or more desirable characters lacked by the other. If the two original parents do not provide all of the desired characters, a third parent can be included by crossing it to one of the hybrid progeny of the first generation (F1). In the pedigree method superior types are selected in successive generations, and a record is maintained of parent–progeny relationships. The F2 generation (progeny of the crossing of two F1 individuals) affords the first opportunity for selection in pedigree programmes. In this generation the emphasis is on the elimination of individuals carrying undesirable major genes.

In the succeeding generations the hybrid condition gives way to pure breeding as a result of natural self–pollination, and families derived from different F2 plants begin to display their unique character. Usually one or two superior plants are selected within each superior family in these generations. By the F5 generation the pure–breeding condition (homozygosity) is extensive, and emphasis shifts almost entirely to selection between families.

The pedigree record is useful in making these eliminations. At this stage each selected family is usually harvested in mass to obtain the larger amounts of seed needed to evaluate families for quantitative characters. This evaluation is usually carried out in plots grown under conditions that simulate commercial planting practice as closely as possible. When the number of families has been reduced to manageable proportions by visual selection, usually by the F7 or F8 generation, precise evaluation for performance and quality begins.

The final evaluation of promising strains involves:

- Observation, usually in a number of years and locations, to detect weaknesses that may not have appeared previously;
- Precise yield testing;
- Quality testing. Many plant breeders test for five years at five representative locations before releasing a new variety for commercial production.

The Bulk–Population Method

The bulk–population method of breeding differs from the pedigree method primarily in the handling of generations following hybridization. The F2 generation is sown at normal commercial planting rates in a large plot. At maturity the crop is harvested in mass, and the seeds are used to establish the next generation in a similar plot. No record of ancestry is kept. During the period of bulk propagation natural selection tends to eliminate plants having poor survival value.

Two types of artificial selection also are often applied:

1. Destruction of plants that carry undesirable major genes
2. Mass techniques such as harvesting when only part of the seeds are mature to select for early maturing plants or the use of screens to select for increased seed size.

Single plant selections are then made and evaluated in the same way as in the pedigree method of breeding. The chief advantage of the bulk population method is that it allows the breeder to handle very large numbers of individuals inexpensively. Often an outstanding variety can be improved by transferring to it some specific desirable character that it lacks. This can be accomplished by first crossing a plant of the superior variety to a plant of the donor variety, which carries the trait in question, and then mating the progeny back to a plant having the genotype of the superior parent. This process is called backcrossing.

After five or six backcrosses the progeny will be hybrid for the character being transferred but like the superior parent for all other genes. Selfing the last backcross generation, coupled with selection, will give some progeny pure breeding for the genes being transferred. The advantages of the backcross method are its rapidity, the small number of plants required, and the predictability of the outcome. A serious disadvantage is that the procedure diminishes the occurrence of chance combinations of genes, which sometimes leads to striking improvements in performance.

Hybrid Varieties

The development of hybrid varieties differs from hybridization. The F1 hybrid of crosses between different genotypes is often much more vigorous than its parents. This hybrid vigour, or heterosis, can be manifested in many ways, including increased rate of growth, greater uniformity, earlier flowering, and increased yield, the last being of greatest importance in agriculture.

CROSS POLLINATED CROPS

The most important methods of breeding cross–pollinated species are:

- Mass selection;
- Development of hybrid varieties;
- Development of synthetic varieties. Since cross–pollinated species are naturally hybrid (heterozygous) for many traits and lose vigour as they become purebred (homozygous), a goal of each of these breeding methods is to preserve or restore heterozygosity.

Mass Selection

Mass selection in cross–pollinated species takes the same form as in self–pollinated species; *i.e.*, a large number of superior appearing plants are selected and harvested in bulk and the seed used to produce the next generation.

Mass selection has proved to be very effective in improving qualitative characters, and, applied over many generations, it is also capable of improving quantitative characters, including yield, despite the low heritability of such characters.

Mass selection has long been a major method of breeding cross–pollinated species, especially in the economically less important species.

Hybrid Varieties

The outstanding example of the exploitation of hybrid vigour through the use of F1 hybrid varieties has been with corn (maize).

The production of a hybrid corn variety involves three steps:

1. The selection of superior plants;
2. Selfing for several generations to produce a series of inbred lines, which although different from each other are each pure–breeding and highly uniform;
3. Crossing selected inbred lines.

During the inbreeding process the vigour of the lines decreases drastically, usually to less than half that of field–pollinated varieties.

Vigour is restored, however, when any two unrelated inbred lines are crossed, and in some cases the F1 hybrids between inbred lines are much superior to open–pollinated varieties. An important consequence of the homozygosity of the inbred lines is that the hybrid between any two inbreds will always be the same.

Once the inbreds that give the best hybrids have been identified, any desired amount of hybrid seed can be produced.

Pollination in corn (maize) is by wind, which blows pollen from the tassels to the styles (silks) that protrude from the tops of the ears. Thus controlled cross–pollination on a field scale can be accomplished economically by interplanting two or three rows of the seed parent inbred with one row of the pollinator inbred and detasselling the former before it sheds pollen. In practice most hybrid corn is produced from "double crosses," in which four inbred lines are first crossed in pairs (A × B and C × D) and then the two F1 hybrids are crossed again (A × B) × (C × D).

The double–cross procedure has the advantage that the commercial F1 seed is produced on the highly productive single cross A × B rather than on a poor-yielding inbred, thus reducing seed costs. In recent years cytoplasmic male sterility, described earlier, has been used to eliminate detasselling of the seed parent, thus providing further economies in producing hybrid seed. Much of the hybrid vigour exhibited by F1 hybrid varieties is lost in the next generation.

Consequently, seed from hybrid varieties is not used for planting stock but the farmer purchases new seed each year from seed companies. Perhaps no other development in the biological sciences has had greater impact on increasing the quantity of food supplies available to the world's population than has the development of hybrid corn (maize). Hybrid varieties in other crops, made possible through the use of male sterility, have also been dramatically successful and it seems likely that use of hybrid varieties will continue to expand in the future.

Synthetic Varieties

A synthetic variety is developed by intercrossing a number of genotypes of known superior combining ability–*i.e.*, genotypes that are known to give superior hybrid performance when crossed in all combinations. (By contrast, a variety developed by mass selection is made up of genotypes bulked together without having undergone preliminary testing to determine their performance in hybrid combination.)

Synthetic varieties are known for their hybrid vigour and for their ability to produce usable seed for succeeding seasons. Because of these advantages, synthetic varieties have become increasingly favoured in the growing of many species, such as the forage crops, in which expense prohibits the development or use of hybrid varieties.

MUTATION BREEDING

Physical Mutagens

Physical mutagens include various types of radiation, viz X–rays, gamma rays, alpha particles, beta particles, fast and thermal (slow) neutrons and ultra violet rays.

X–Rays

X–rays were first discovered by Roentgen in 1895. The wavelengths of X–rays vary from 10–11 to 10–7. They are sparsely ionizing and highly penetrating. They are generated in X–rays machines. X–rays can break chromosomes and produce all types of mutations in nucleotides, *viz.* addition, deletion, inversion, transposition, transitions and transversions. X–rays were first used by Muller in 1927 for induction of mutations in Drosophila. In plants, Stadler in 1928 first used X–rays for induction of mutations in barley.

Gamma Rays

Gamma rays have shorter wave length than X–rays and are more penetrating than gamma rays. They are generated from radioactive decay of some elements like 14C, 60Co, radium etc. Of these, cobalt 60 is commonly used for the production of Gamma rays. Gamma rays cause chromosomal and gene mutations like X–rays.

CHEMICAL MUTAGENS

The chemical mutagens can be divided into four groups, viz.

1. Alkylating agents,
2. Base analogues,
3. Acridine dyes,
4. Others. A brief description of some commonly used chemicals of these groups is presented below.

Some Commonly Used ChemicalMutagens and Their Mode of Action

Group of mutagen	*Name of chemical*	*Mode of action*
1.Alkylating Agents	Ethyl methane	↔AT GC Transitions
	Sulphonate Methyl	Transitions
	Methane Sulphonate	↔GC AT Transitions
	Ethyl Ethane	Transitions
	Sulphonate	
	Ethylene Imines	
1.Base	5 Bromo Uracil	↔AT GC Transitions
	2 Amino purine	↔AT GC Transitions
Analogues	Acriflavin, Proflavin	Deletion, addition and frame shifts.
1.Acridine Dyes		
1.Others	Nitrous Acid	↔AT GC Transitions
	Hydroxylamine	↔GC AT Transitions
	Sodium Azide	Transitions

The speed of hydrolysis of the chemical mutagens is usually measured by the half life of the chemicals. Half life is the time required for disappearance of the half of the initial amount of active reaction agent. The following table gives the half life in hours at different temperatures.

In the case of DES the mutagenic solution should be changed at every half an hour to get good results. Half life is the function of temperature and pH for a particular compound. One should be extremely careful in handling alkylating agents since most of them are carcinogenic.

Especially for ethylene imine, it should be handled under aerated conditions.

EMS though not dangerous, it should not be pipetted out by mouth. Besides the alkylating agents, we are also having chemical mutagens like, Base analogues, Acridine dyes, Antibiotics and other miscellaneous chemicals.

Treatment of Seeds with Mutagenic Chemicals

- *Materials Required:* Conical flask, beaker, pipette, glass rods, measuring cylinder, stop watch, distilled water and phosphate buffer.
- *Method:* Mutagenic chemical is diluted to the required concentration by using distilled water. To prepare the molar concentration of DES, the method is

$$\frac{\text{Molecular weight} \times \text{a.i. (purity percentage)}}{\text{Specific gravity (active ingredient)}}$$

$$\text{Eg.DES} = \frac{154}{1.18} \times \frac{100}{99} = 131.\text{CC}$$

131 CC dissolved in one litre will give 1 molar solution.

Seeds have to be soaked in the distilled water for different hours depending upon the seeds, to initiate biochemical reactions.

The chemical action is found to be affected by the frequency and spectrum of mutagen depending upon the stage of cell division, during the process of germination. If the chemical treatment is synchronized with DNA synthesis stage (G1, S and G2) then we can get better results.

The presoaked seeds are taken in a flask and chemical is added. Usually the quantity of the chemical is ten times the volume of seeds. Intermittent shaking should be given to ensure uniform exposure of the chemicals. The chemical should be drained after the treatment time is over. The seeds should be washed thoroughly in running tap water, immediately for not less than 30 minutes. After washing, the seeds should be dried in between the filter paper folds. Seeds are to be arranged in germination tray with equal spacing. Trays are kept in a controlled environment of temperature and humity. Periodical observation on germination upto 10–15 days is needed. From the germination percentage, we can assess the LD50 dose.

PARTICIPATORY PLANT BREEDING

Certain trends have made the world ripe for adoption of participatory plant breeding methods.

- The shielding of crops from environmental stress in high-input systems is facing increasing economic and ecological problems. Scientists are changing their attitudes, and a new paradigm is being formulated. Instead of modifying the environment to suit the requirement of high yielding varieties, the varieties need to be modified to suit the environment.
- The claim that modern varieties can be made broadly adapted and be superior across most farming environments within an ecogeographic region is being challenged.

- If relevant diversity exists in a locale, the combined action of natural and artificial selection within a local environment may be an efficient breeding method. Experiments show that this may work also in a fertilizer-intensive system.
- In recent years farmer groups working with local seeds have been organized all over the world. They are not primarily conservers of old seeds. They want their seeds to be improved, in an evolutionary, slow and steady way, and under their own control.
- Finally, participatory methods have been developed in order to facilitate the involvement of farmers together with scientists as active and equal partners in research to generate relevant farm technology. Such methods can be applied also to plant breeding.

Commercial seeds often diffuse into areas where the traditional seed supply system is still predominant. Farmers try them with an open mind and adopt or reject them according to their own criteria. If grown and multiplied in the villages, diversity will start to appear within them and local reselection will be possible. In that way commercial varieties eventually might become like landraces. It is also commonly observed that farmers change and exchange seeds. A traditional farming system rarely functions as an environment for the static preservation of old landraces.

Often practices range from neglect to very simple mass selection, but with a few scattered individuals who devote an exceptional amount of effort to the maintenance or improvement of seed quality. These exceptional persons, very often women, may be the source of good seeds for others in the community. Once such a community is organized for seed management and improvement, it becomes possible for it to establish links to scientific institutions.

Community resources for participatory approaches to seed management and breeding are not limited to indigenous culture and traditional practices. The educational status and experiences of modern farmers may also be turned into a resource for community action. In the Philippine group, a number of the members were high school graduates and a few had a university degree. And moreover, most of them had a couple of decades' experience with modern input-intensive farming. Seed activities opened their minds towards the traditional societies, towards themselves and towards the modern world. The traditional societies supplied them with their seeds, and through the seeds, they learned to appreciate the values and achievements of these societies. They discovered their own potential, and saw that the outside world could bring more than technology packets: it could bring knowledge and ideas to be exploited and further developed by themselves.

CONVENTIONAL BREEDING

Conventional plant breeding involves crossing carefully chosen parent plants, then selecting the best plants from the resulting offspring to be grown on for further selection. For cereals, hundreds of individual crosses are carried out by hand to create seed for the first filial (or F1) generation.

The resulting F1 plants are uniform, but in the following generation several hundred thousand different plants are produced. Because of the way genes work, the new combinations produced from each cross are not revealed until the second (F2) generation.

It is this enormous diversity of new gene combinations that may hold the key to a successful new variety. The plant breeder's task is to select the plants most likely to meet his breeding objectives. Seed from the best of these F2 plants is grown on in small rows or plots and the best plants again selected–this process is repeated year after year until only the very best plants remain. As promising new lines emerge, tests are conducted on each plot to assess factors such as yield, disease resistance and endues quality.

Once the best lines are purified to ensure that every plant has the same characteristics, the process of multiplying seed begins. These 'inbred' lines are then ready for entry into official trials, some six to ten years after the initial cross.

Hybrid Breeding

For some crop species, the seed supplied to growers is that produced from the first cross between selected parents. The resulting varieties, known as F1 hybrids, offer potential advantages in crop performance. Hybrid breeding is widely used to produce varieties of field vegetables, maize and oilseed rape. F1 hybrids are unique in expressing 'hybrid vigour' in the growing crops for a single year. This may result in higher yields, greater uniformity, or improvements in quality. Unlike inbred lines, F1 hybrids do not breed true year after year and their performance gains are not maintained in subsequent generations.

Improved Breeding

With increasing knowledge and improved technology, breeders have developed ways to enhance the speed, accuracy and scope of the breeding process.

There are several ways to reduce the lengthy interval between the first cross of selected parents and establishing true breeding lines of promising new varieties. For example, maintaining parallel selection programmes in northern and southern hemispheres allows two generations to be produced each year. Single seed descent produces very small plants under restricted growth conditions. Large numbers of these plants are cultivated in artificial growth rooms, with two or more generations produced in a year. In potatoes, mini-tuber breeding speeds up the slow multiplication process by producing miniature plants under greenhouse conditions. More recent laboratory techniques enable breeders to operate at the level of individual cells and their chromosomes. In certain crop species, such as potatoes and oilseed rape, it is possible to produce new varieties in the laboratory through

Protoplast Fusion

Individual plant cells with their outer walls removed are fused, and the fused cells then induced to divide and grow in a culture medium. Whole plants are eventually regenerated containing new combinations of genes from the two parents.

Embryo Rescue and Assisted Pollination

Allow breeders to expand the range of available characters by making crosses between species, which would not produce viable offspring outside the laboratory.

Double Haploid Breeding

Enables breeders to produce genetically uniform lines within one generation. This effectively by-passes the lengthy process of self pollination and selection normally required to produce true breeding plants. Latest developments in genetic science have greatly improved our understanding of how plants behave, offering additional ways to enhance the breeding process.

Genomics

By mapping the genetic makeup, or genome, of crop species, scientists can identify the exact position and function of individual genes. Genome mapping has revealed striking similarities in the genomes of different crop species, such as rice, wheat, barley and rye. This information is already helping to broaden the scope and precision of current breeding programmes.

Marker Assisted Breeding

Allows breeders to determine whether desired traits are present in a new variety at an early stage in the breeding programme.

Genetic Modification

Genetic modification of plants is achieved by adding a specific gene or genes to a plant, or by knocking out a gene with RNAi, to produce a desirable phenotype. The resulting plants are often referred to as transgenic plants. Genetic modification can produce a plant with the desired trait or traits faster than classical breeding because the majority of the plant's genome is not altered.

To genetically modify a plant, a genetic construct must be designed so that the gene to be added or knocked-out will be expressed by the plant. To do this, a promoter to drive transcription and a termination sequence to stop transcription of the new gene, and the gene of genes of interest must be introduced to the plant.

A marker for the selection of transformed plants is also included. In the laboratory, antibiotic resistance is a commonly used marker: plants that have been successfully transformed will grow on media containing antibiotics; plants that have not been transformed will die. In some instances markers for selection are removed by backcrossing with the parent plant prior to commercial release.

The construct can be inserted in the plant genome by genetic recombination using the bacteria Agrobacterium tumefaciens or *A. rhizogenes*, or by direct methods like the gene gun or microinjection. Using plant viruses to insert genetic constructs into plants is also a possibility, but the technique is limited by the host range of the virus. For example, Cauliflower mosaic virus (CaMV) only infects cauliflower and related species. Another limitation of viral vectors is that the virus is not usually passed on the progeny, so every plant has to be inoculated.

The majority of commercially released transgenic plants, are currently limited to plants that have introduced resistance to insect pests and herbicides. Insect resistance is achieved through incorporation of a gene from Bacillus thuringiensis (Bt) that encodes a protein that is toxic to some insects. For example, if cotton pest the cotton bollworm feeds on Bt cotton it will ingest the toxin and die. Herbicides usually work by binding to certain plant enzymes and inhibiting their action. The enzymes that the herbicide inhibits are known as the herbicides *target site*. Herbicide resistance can be engineered into crops by expressing a version of *target site* protein that is not inhibited by the herbicide. This is the method used to produce glyphosate resistant crop plants.Genetic modification of plants that can produce pharmaceuticals (and industrial chemicals), sometimes called *pharmacrops*, is a rather radical new area of plant breeding

Proteomics

Allows breeders to understand how genes behave in different parts of the plant and under different growing conditions.

Maintaining Genetic Diversity

Maintaining biodiversity is central to the process of crop improvement. It is in every breeder's interest to ensure that the gene pool from which new traits are selected remains as extensive as possible. Plant breeders created the first gene banks in the 1930s to conserve the valuable genetic diversity within past and present varieties, as well as landraces and wild relatives of cultivated crop species. Plant breeding is integral to ongoing initiatives to identify, classify and conserve existing biodiversity.

Testing Plant Varieties

Before any new crop variety can be placed on the market, it must undergo statutory testing under a process known as National Listing. Successful varieties are placed on a National List or register of varieties approved for marketing.

National Listing rules are determined on a European basis, and apply to all the major agricultural and vegetable crop species. Official trials are conducted, in most cases for a minimum of two years, to test each new variety for a range of characteristics which together determine its uniqueness, its genetic uniformity, and its value to growers and the rest of the food chain. National Listing is extremely rigorous–the majority of varieties entered do not complete the process. In winter wheat, for example, only a quarter of varieties entered for National Listing during the 1990s were finally approved.

National Listing–DUS

All varieties entered for National Listing are assessed for Distinctness, Uniformity and Stability (DUS). In the case of cereals, some 30 individual characteristics of the plant are inspected to verify that it is distinct, ie clearly distinguishable from other varieties, that its characteristics are uniform from one plant to another, and that the variety is stable in that it breeds true to type from one generation to the next.

National Listing–VCU

For agricultural crops, National Listing also involves trials to establish a variety's Value for Cultivation and Use (VCU). This provides an assurance to growers that only varieties with improved performance or end-use quality can be officially approved for sale.

Recommended and Descriptive Lists

Once a new variety has been added to the National List, it is cleared for marketing. However, its success in the market place is by no means guaranteed. Further independent trials are conducted each year to compare the performance and quality of the best varieties. These trials provide the basis for detailed information and advice to growers and their customers. As shown in the following illustration for winter wheat, the process of statutory and commercial evaluation can take up to five years. Only a handful of varieties clear all these hurdles, which are in addition to the many years spent testing and selecting in the breeder's own trials programme.

Genetically Modified Varieties

No GM crops can be marketed until they have been assessed and approved in terms of human health, food safety and the environment. The cross-border movement of genetically modified organisms (GMOs) is regulated at a global level under the internationally agreed Biosafety Protocol.

Variety Maintenance and Identity Preservation

As well as developing new varieties, plant breeders maintain the genetic purity of existing lines and pre-commercial seed supplies year by year. This process is costly and time-consuming, but essential to maintain the quality and performance of each variety.

For cereals, variety maintenance begins after a few years of selection trials, when the promise of a variety is just emerging. At that stage, all that exists of what may become a widely grown variety is a single row containing around 100 plants. The breeder then bulks up supplies of the purified lines of breeder's seed into prebasic and then basic seed. Each year specialist seed growers are used to grow basic seed for the first generation of certified seed– or C1 seed. After one more year this becomes C2 seed, the main source of certified seed used by farmers.

The plant breeder continuously maintains breeder's seed for the process of multiplication through pre-basic, basic, C1 and C2 seed to ensure the variety's performance and quality year after year. Greater emphasis is now being placed on preserving the identity of individual varieties after harvest, both to conserve quality characteristics and to meet consumer demands for assurances about the integrity and traceability of their food.

Seed Certification

Seed of an approved variety can only be marketed if it meets strict quality criteria Seed quality standards are laid down in UK and EU law, and policed by agencies appointed by Government. The UK's official seed certification system offers an independent assurance of quality to growers. Minimum standards apply for varietal identity, purity and germination capacity. In addition, strict limits apply to seed-borne diseases, and the presence of physical impurities such as weed seeds. Around 9% of the UK arable area is used to multiply the pure lines of seed from the plant breeder into certified seed.

Several thousand individual crops are involved; each grown under specific management regimes to ensure the purity and integrity of the resulting seed is maintained. To gain certification, every seed crop must undergo crop inspection and seed testing. Seed certification underpins the health and purity status of the major arable crops in Britain. It offers an independent benchmark of quality on which buyers of seed and their customers depend.

Farm-saved Seed

For certain crop species–particularly small-grain cereals–growers can opt to save their own seed for sowing the following year provided care is taken to ensure that the crop remains healthy and free from impurities, and that the resulting seed is carefully conditioned and cleaned. Without independent testing for germination and freedom from seed borne diseases, however, farm-saved seed can harbors risks to growers that may not become apparent until well into the growing season. Many growers choose certified seed for the peace of mind that quality and performance is independently assured.

PLANT BREEDING IN THE FUTURE

Success in any line of business depends on being able to anticipate and respond to changing market requirements. Such is the long-term nature of developing a new crop variety that plant breeders are in the unenviable position of forecasting their customers' demands at least five, and probably nearer ten years, in advance. Plant breeding objectives will continue to be shaped by changes in agricultural and environmental policy. There is no doubt that while improvements in crop yield and endues quality remain of paramount importance, the future for British agriculture lies in crop production systems which deliver both economic and environmental benefits.

Increasingly, plant breeders are targeting varieties suitable for low input and organic regimes. It is already estimated that genetic improvements in disease resistance alone save British farmers more than £100 million a year in reduced agrochemical use and improved yield. A range of other factors also determines a variety's suitability for low input, integrated farming systems. They include the crop growth cycle, straw strength and susceptibility to weed competition, as well as the variety's ability to thrive in a range of soil types, climatic conditions and rotational patterns.

Bibliography

Alfred Byrd Graf: *Advances in Plant Physiology*, Rajat Pub, Delhi, 2008.

Bandyopadhyay, P C : *Breeding and Crop Production*, Gene Tech Books, Delhi, 2007.

Barnett, S A : *Principles of Ethology and Behavioural Physiology*, J V Pub, Delhi, 2006.

Boyer , J.S.: *Measuring the Water Status of Plants and Soils*, Academic Press, N.Y., 1995.

Chambers, R.: *Rural Development: Putting the Last First*, London, Longman, 1983.

Clark, R B : *Physiology of Crop Production*, International Book, Delhi, 2007.

Collymore L.: *Fruit Production in Barbados*, Port of Spain, Trinidad and Tobago, 1996.

Coste R.: *Coffee: the Plant and the Product*, London, MacMillan, 1992.

Crucefix D.: *Avocado Variety Selection for Export Development*, Roseau, CARDI, 1996.

Dabholkar, A.R. : *General Plant Breeding*, Concept, Delhi, 2006.

Daubenmire, R. F.: *Plants and Environment*, New York, Wiley, 1947.

David White: *The Physiology and Biochemistry of Prokaryotes*, Oxford University Press, Delhi, 2007.

Dharamvir Hota: *Modern Biotechnology in Plant Breeding*, Gene Tech Books, Delhi, 2007.

Dinabandhu Sahoo: *Farming the Ocean : Seaweeds Cultivation and Utilization*, Aravali, Delhi, 2000.

Doijode S. D.: *Seed Germination in Fruits*, New Delhi, Malhotra Publishers, 1993.

Ferentinos L.: *Proceeding of the Sustainable Taro Culture for the Pacific Conference*, Honolulu, HITAHR, 1993.

Georges, S.: *The Debt Boomerang: How Third World Debt Harms Us All*, Boulder, Westview Press, 1992.

Gowen S.: *Banana and Plantains*, London, Chapman Hall, 1996.

Hardy B.: *Biology and Agronomy of Forage Arachis*, Cali, International Centre for Tropical Agriculture, 1994.

Herminie Broedel Kitchen: *Soils and Crops : Diagnostic Techniques*, Satish Serial Publishing, Allahabad, 2004.

Jain, Sumit : *Biotech's Dictionary of Plant Breeding and Genetics*, Biotech, Delhi, 2005.

Jeffers P.: *Evaluation of Four Onion Varieties in Montserrat*, Plymouth, CARDI, 1992.

Jones, R. M.: *Plant Resources of South-East Asia*, Wageningen, Pudoc Scientific Publishers, 1992.

Kannaiyan, S. : *Rice Management Biotechnology*, Associated, Delhi, 1995.

Kataria, T N : *Plant and Crop Physiology*, Pearl Books, Delhi, 2008.

Keshav Prasad Yadav: *Application of Morphometry in Geomorphology*, Radha Pub, Delhi, 2008.

Khan, S.: *Plant Breeding Advances and in vitro Culture*, CBS, Delhi, 1997.

Kumar, N. : *Breeding of Horticultural Crops : Principles and Practices*, New India Pub, Delhi, 2006.

Kunelius T.: *Annual Ryegrasses in Atlantic Canada*, Ottawa, Agriculture Canada, 1991.

Lal. Madan : *Physiology, Biochemistry and Biotechnology*, Manglam Pub, Delhi, 2007.

Lorenzen, S.: *The Phylogenetic Systematics of Freeliving Nematodes*, London, The Ray Society, 1994.

Madan Lal Bagdi: *Physiology, Biochemistry and Biotechnology*, Manglam Pub, Delhi, 2007.

Mishra, U K : *Physiology of Animal Growth and Bioenergetics*, Satish Serial Pub House, Delhi, 2008.

Mitra S.: *Postharvest Physiology and Storage of Tropical and Subtropical Fruits*, Oxon, CABI, 1997.

Nobel, P. S.: *Physicochemical and Environmental Plant Physiology*, Academic Press, San Diego, 1999.

Pareek O. P.: *Advances in Horticulture: Fruit Crops*, New Delhi, Malhotra Publishing House, 1993.

Pemberton, R. W.: *Predictable Risk to Native Plants in Weed Biological Control*, Oecologia, 2000.

Politycka, A.R. and C L Goswami: *Plant Physiology : Research Methods*, Scientific, Delhi, 2007.

Rajni Sharma: *An Introduction to Plant Morphology*, Campus, Delhi, 2004.

Rao, P. Venkateshwara : *Dairy Farm Business Management*, Biotech Books, Delhi, 2008.

Saini, M.L. : *Plant Breeding and Crop Improvement*, CBS, Delhi, 1997.

Samiullah Khan: *Plant Breeding Advances and in vitro Culture*, CBS, Delhi, 1997.

Sathe, T V : *Biotechnological Approaches in Entomology*, Manglam Pub, Delhi, 2008.

Sharma, R.C. : *Diseases of Horticultural Crops: Fruits*, Indus, Delhi, 1999.

Sharma, Rajni : *An Introduction to Plant Morphology*, Campus, Delhi, 2004.

Shivanand Tolanur: *Practical Soil Science and Agricultural Chemistry*, International Book Distributing, Delhi, 2004.

Shubhrata R. Mishra: *Morphology of Plants*, Discovery, Delhi, 2004.

Singh, S K : *Biotechnology, Plant Propagation and Plant Breeding*, Campus Books, Delhi, 2008.

Subhash Ranade and Sunanda Ranade: *Concept of Ayurvedic Physiology*, Narendra Prakashan, Delhi, 2003.

Tripathi, Pramila : *Botanical Pesticides in the Management of Post-harvest Fruit Diseases*, Daya, Delhi, 2005.

Tyagi, I.D. : *Plant Breeding and Genetics at a Glance*, South Asian, Delhi, 2005.

Vanderplank, J.E.: *Plant Diseases: Epidemics and Control*, New York, NY, USA, Academic Press, 1963.

Verma, L.R. and R.C. Sharma: *Diseases of Horticultural Crops: Fruits*, Indus, Delhi, 1999.

Index

A

Agriculture Sector, 15.
Agrobacterium-mediated, 62, 66, 67.
Anatomy of Plant Cell, 98.

B

Backcross Method, 209.
Beneficial Traits, 7, 183.
Bioinformatics, 15.
Biological Cell, 31, 37.
Biological Energy, 33.
Biological Viruses, 35.
Building with Silk, 179.

C

Carbohydrates, 77, 98, 107.
Cell Respiration, 34.
Cheese Chymosin, 173.
Chemical Composition, 91.
Consumer Traits, 196.
Controlled Ripening, 175.
Crop Plants, 6, 16, 41, 75, 137, 153, 183, 184, 185, 186, 187, 189, 199, 200, 217.
Crop Rotation, 193.
Cross Pollinated Crops, 209.

D

Discovery of Transposition, 71.

E

Enkephalins, 153, 188.
Environmental Concerns, 158.
Escherichia Coli, 180.
Eukaryotic Cells, 79, 94.
Evolution in Vascular Plants, 113.

F

Food Safety, 6, 152, 156, 159, 160, 165, 200, 219.

G

Gene Mapping, 50.
Gene Silencing, 68, 69.
Gene Testing, 24.
Genetic Diversity, 58, 114, 118, 130, 131, 132, 147, 195, 196, 218.
Genetic Engineering, 7, 12, 14, 18, 39, 59, 68, 159, 160, 166, 167, 168, 169, 171, 172, 174, 179, 181, 182, 185, 190, 191, 194, 195, 196, 197, 200, 204.
Genetic Variations in Plants, 128.
Genome Evolution in Plants, 70.
Glycosides, 157.

H

Herbicides, 193.
Hermocellum, 186.

I

Immunoglobulin, 187.

K

Kanamycin, 161.

L

Lawns, 197.
Legitimate Concerns, 155.
Living Species, 105.

M

Medicinal Discovery, 15.
Medicinal Research, 15.
Metabolism, 79, 81, 82, 96, 97.
Microorganism, 150, 151.
Minerals, 136.
Misinformation, 163, 164, 166.
Mode of Reproduction, 19, 137, 138, 147.
Mothers for Natural Law, 166.
Mutation Breeding, 211.

N

Nutrient, 16, 143, 144, 159, 167.

P

Pesticide Residues, 168, 200.
Phenotypic Variation, 119, 129.
Phosphinothricin, 149, 150.
Plant Breeding, 1, 7, 19, 138, 151, 192, 193, 194, 195, 197, 198, 199, 200, 203, 204, 213, 214, 215, 217, 218, 220.
Plant Cell Structure, 77.
Plant Genetic Engineering, 18.
Plant Genetics, 19.
Plant Reproduction, 105, 113, 125, 139.
Plant Reproduction System, 125.
Plant Transposons, 64, 75.
Polygalacturonase, 152.
Polymer Chemistry, 45.

R

Raw Materials, 33.
Recombinant Proteins, 149.
Reproduction in Plants, 105.
Reproductive Systems, 114, 115, 120, 147, 200.
Restriction Enzymes, 169.
Reticulum, 186.
Rhizosecretion, 186.
RNA Interference, 28.
Rust Resistance, 195.

S

Safer Meat, 180.
Selection Methods, 137, 139.
Self Fertilizing Crops, 138, 139.
Sustainable Growth, 16.
Sweet Potato, 16.

T

Transgenic Plants, 5, 57, 149, 168, 180, 181, 182, 183, 184, 186, 187, 188, 189, 190, 191, 217.

V

Vascular Plants, 77, 78, 97, 105, 112, 115, 116, 133, 134, 147.
Vitamins, 3, 164.

W

Witchweed Control, 172.

❑❑❑